Geography 11 - 16

Published in 1995, this book seeks to show how geography teachers can meet the requirements of the National Curriculum at Key Stages 3 and 4 without abandoning principles of good practice. It considers geographical education in the contexts of geography itself, society and education. Deriving principles of good practice from these contexts, the author gives guidance on how to produce case studies (or curriculum units) that both reflect these principles and respond to the requirements of the National Curriculum following the Dearing Revised Order.

Geography 11 - 16

Rekindling Good Practice

Bill Marsden

Routledge
Taylor & Francis Group

First published in 1995
by David Fulton Publishers Ltd

This edition first published in 2018 by Routledge
2 Park Square, Milton Park, Abingdon, Oxon, OX14 4RN
and by Routledge
711 Third Avenue, New York, NY 10017

Routledge is an imprint of the Taylor & Francis Group, an informa business

Publisher's Note
The publisher has gone to great lengths to ensure the quality of this reprint but points out that some imperfections in the original copies may be apparent.

Disclaimer
The publisher has made every effort to trace copyright holders and welcomes correspondence from those they have been unable to contact.

A Library of Congress record exists under LCCN: 95201700

ISBN 13: 978-1-138-48960-8 (hbk)
ISBN 13: 978-1-351-03270-4 (ebk)
ISBN 13: 978-1-138-48961-5 (pbk)

GEOGRAPHY 11–16

Rekindling Good Practice

Bill Marsden

David Fulton Publishers

London

David Fulton Publishers Ltd
2 Barbon Close, London WC1N 3JX

First published in Great Britain by
David Fulton Publishers 1995

British Library Cataloguing in Publication Data

A catalogue record for this book is available from the British Library

ISBN 1–85346–296–9

Typeset by Harrington & Co.

Contents

Acknowledgements iv

Series Editor's Foreword..v

Preface vii

  1 Introduction: Theory, Practice and Educational Aims....................1

Section A: GEOGRAPHY AND GEOGRAPHICAL EDUCATION 12

  2 The Changing Nature of Geography...12

  3 Continuity and Change in Geographical Education 28

  4 Perspectives on Geography and Curriculum Integration.............42

Section B: GEOGRAPHY AND EDUCATION....................................59

  5 Curriculum Theory and Geography...59

  6 Teaching and Learning Geography..78

  7 Assessment, Testing and Examinations 99

Section C: GEOGRAPHY AND SOCIAL EDUCATION 119

  8 Geographical Education and Social Stereotyping.......................119

  9 The Values and Attitudes Dimension: Issues-based Geography 137

Section D: NATIONAL CURRICULUM PLANNING IN

GEOGRAPHY 154

  10 National Curriculum Geography in its Cross-curricular Context 154

  11 Geography in the National Curriculum 167

  12 National Curriculum Planning at Key Stage 3 181

  13 W(h)ither Geography 14–16? 195

CONCLUSION 206

  14 Approaching the Millennium..206

  References..216

  Index..227

Acknowledgements

Acknowledgement is made for material in or for drawing the following illustrations:

Figure 2.1, based on W.Kirk (1963).
Figure 2.2, based on R.J. Johnston (1991).
Figures 4.1, 4.2, 5.2., 5.3, 5.4, 6.1, 6.2 and 8.1, drawn or redrawn by Windowgraphics, Southport.
Figures 4.3, 4.4, and 4.5, from C.S. Tann and S. Catling (1988).
Figure 5.2, from D.J. Wheeler (1967).
Figure 5.4, from Schools Council (1972) in R. Pring (1989).
Figure 8.1, from E.D. Mapother (1870).
Figure 8.2, from W.E.and V.M. Marsden (1986), *World Concerns.* Edinburgh: Oliver and Boyd.
Figure 9.1, from W.E. and V.M. Marsden (1986) *Britain.* Edinburgh: Oliver and Boyd

Preface

Since the passing of the Education Reform Act of 1988, many in the world of education have argued that one of the failings of the National Curriculum has been the neglect of principles of good educational practice. Perhaps this is not surprising when the working context is one of intolerant polarisation: of a government- and media-generated climate of opinion celebrating their 'good 'e's' (economic expediency, efficiency, effectiveness, elitism, and examinations) and deriding the 'bad 'e's' (explicitly, educationists, 'experts', and 'enthusiasts' and, implicitly, concepts of emancipation, enlightenment and equity). To those in the former camp, good practice means a return to the so-called basics: a revival of formal instructional practice. In this book, back to basics implies achievement of a balance: between good practice in geography as a subject, good educational practice, and good social education, defined here as including environmental education. It is a balance that has rarely if ever been realised because, I would argue, during each period of past change, we have tended to concentrate too much on one of these key variables. Thus in the grammar school tradition, the subject variable was over-emphasised; in the progressive primary tradition, the educational variable; and in the comprehensive school tradition, the social. The imbalance remains.

We are the inheritors, not the inventors of good practice. We must not see ourselves as the beneficiaries of unchecked progress from the perceived primitivism of earlier times to a current superior state. As we know only too well from the contemporary situation, the forces of reaction and progress are constantly present and in a state of tension. Even in the most reactionary circumstances, the embers of good practice have continued to flicker. Our educational heritage, with all its strengths and weaknesses, is there for us to draw upon. There have been conservers, misusers and, mostly, ignorers of this heritage. As we shall find, hardly any of the dramatic developments in geographical education over the last 30 years have not been anticipated in previous periods. We have invented few, if any, new wheels, and what is worrying is that many seem to think they have. What might be a new wheel, however, would be the achievement of a better balance between the elements of good practice cited above.

Rawling has drawn attention to the neglect of key variables and general lack of balance in curriculum development in recent times: 'It is essential for the future well-being of the subject in schools that the nature of geography as a school subject is first debated, and adherence to certain principles agreed so that we can face the future with confidence' (1992, p.292). Similarly '...the 7 years 1993–2000 have the potential to become the 7 more balanced years for school geography, avoiding both the over-indulgent confidence of 1970–77 and the narrow political emphasis and impoverished curriculum thinking of the 1980–87 period' (1993, p.116).

The impoverishment in the last 15 years or so has indeed been chiefly in the geographical and educational (curriculum theory and pedagogy) variables. Uncritical implementation of the National Curriculum could lead us into another imbalance. The educational and geographical literature of the 1960s and 1970s led me, in *Evaluating the Geography Curriculum* (Marsden, 1976), over-optimistically to believe we were on the road to a new and balanced synthesis. What strikes me now is that, in common with much other educational writing of its time, that book was so detached from any sense of a brewing political ferment. The *Black Papers* might never have been written. This complacency was soon to be shattered by the external forces to be described in later chapters. Re-reading the older literature on which *Evaluating the Geography Curriculum* was built, however, also renews the impression that we then had richer resources in various educational disciplines, as well as in geography, to draw upon, than we do today. But as the more recent store of thought has hardly as yet been raided, in our preoccupation with the National Curriculum, part of the purpose of this book is to investigate whether there is a current resource of fundamental thinking in geography and education that we can make use of in a revitalised critique. My cautious conclusion is that there is, and that it complements rather than supplants earlier thinking. Much in the historic heritage needs to be conserved: that is why parts of *Evaluating the Geography Curriculum* are redeployed here. A palimpsest approach is used: parts of the old slate have to be wiped clean, but others come through as perennially useful, even while new and potentially fruitful thinking appears. Such an approach seems to me vital in achieving a new balance and in rekindling good curriculum practice.

Readers of this book had better be warned that it is written from a liberal–humanist, left-of-centre standpoint. Those who are bored or repelled by arguments for compromise and consensus, and who look for ideological purity rather than eclecticism and pluralism, would be better advised to seek out more *ipse dixit* literature. An important objective is, however, to maintain a critical stance, not in the sense of a 'radical critique' but in the sense of an 'intellectual critique', while keeping in the forefront values issues. Thus the National Curriculum will be subjected to what I hope is an intellectual critique but not to an uncritical dismissal. Because it has fundamental flaws does not mean that it does not also provide opportunities. If professional teachers wish to seek the balance of good practice being advocated, the National Curriculum will not stop them.

This book will therefore be structured round the three key variables, to which we need to attach our search for good practice.

First there is a survey of relevant literature on educational aims which must provide the foundation of worthwhile educational practice.

Secondly, geography will be considered in its subject context, covering its changing nature at the frontiers of research, its justification in the school curriculum, and its links with other parts of that curriculum.

Geography in its educational context, exploring aspects of curriculum theory, research on children's thinking, relevant pedagogy, the role of the teacher, and assessment and testing, will form the third section of the book. Then geography in its social context will be investigated, covering the problems of social stereotyping, the values and attitudes dimension, and issues-based geography.

In the final section, these and other matters will be considered in the context of the appearance of the National Curriculum in geography: how the National Curriculum came about; changing notions about what should be its content; some guidance on curriculum planning at Key Stage 3; and a brief chapter on developments in the 14–16 age range. It might be mentioned here that the timing of change in relation particularly to post-Dearing assessment arrangements and Key Stage 4 issues has sat awkwardly with the target period for the completion of this book. The attempt has been made to keep it as up-to-date as possible.

The last thing that is intended here is any attempt to offer teacher-proof solutions. The book should be seen as a guide charting pathways through the undergrowth of a wide and complex range of variables, that might be difficult for the hard-pressed classroom teacher to find time to carve out at first hand. It is hoped that the pathways will lead close to the destination of practical resolutions. But these can only satisfactorily be achieved in the final choices and synthesis of the professional practitioner.

CHAPTER 1

Introduction: Theory, Practice and Educational Aims

Theory and Practice

'In other words, theory is practice become conscious of itself, and practice is realised theory', concluded James Welton, Professor of Education in the University of Leeds, in the early years of this century. Without such theory, the 'mechanical teacher does as he was done by; with him progress implies change, and change is unwelcome, for he cannot adapt himself to it.' At the same time, the theory should not be 'an unsubstantial vision spun out of the clouds of an untrammelled imagination...a theory of teaching which deserves the name is in the closest possible touch with school work' (1906, pp.17–18).

How far the pedagogues of the time drew upon this maxim is open to speculation. Generations of teachers have on the face of it been suspicious of educational theory. One of the tensions that must be accepted is that theory does not make life easier for the teacher by providing a set of recipes. Far from ensuring successful practice, it draws attention to the complex nature of educational issues. It can only present generalisations, and the practitioner must exercise skilled judgement as to how to apply them in particular situations (Entwistle, 1971, p.101). As Knight has put it, 'research does not provide codes for practitioners to follow, but it does produce healthy food for thought' going 'hand-in-hand with teacher professionalism' (1993, p.3).

Implicit in this discussion is therefore the distinction between the rule-obeying 'craft teacher', and the more autonomous and creative 'professional teacher', critically reflective of what s(he) is doing. Part of the broader framework the professional will address is the theoretical basis of practice, for such is needed to be able to examine and re-examine what is meant by good practice, and how this might continue or change over time. This is particularly vital in a context in which the implementation of the National Curriculum by narrow 'craft teachers' could well reinforce the limited utilitarian thinking behind the official line.

The distinction between 'craft' and 'professional' teachers must not be polarised (Daugherty and Lambert, 1994, pp. 342–4). In the real world most teachers will in their practice reflect elements of both traditions. In the current situation, an increasing number of tasks demand bureaucratic skills. Being a good crafts-person is in any case not dishonourable. Whether or not teachers are fundamentally hostile to theory is also a moot point. Clearly if asked the question directly, they will be likely to indicate that they have more pressing demands to meet than perusing the research literature. The issue of whether teachers are influenced by theory, however, is broader and more important. As will be suggested later, there is no doubt in my mind that developments in curriculum theory have had, since they revolutionised educational thinking in this country in the 1960s, a profound impact on classroom practice. The stance taken here is that there is an essential inter-penetration between theory and practice, and that recognition of this cross-fertilisation is a key criterion in recognising the truly professional teacher. Schools do recognise they must have aims, and the first aspect of the resource of theory to be considered, not surprisingly, is about these.

Educational Aims

As White (1990) has indicated, the National Curriculum is extremely rich in its presentation of specific objectives, but poverty-stricken in its attention to aims. This does not, of course, mean that the government has no aims for the national system of education: rather, perhaps, that it prefers that they should remain undebated. It merely offers the bland and unchallengeable rhetoric that the school must 'promote the spiritual, moral, cultural, mental and physical development of pupils', and 'prepare such pupils for the opportunities, responsibilities and experiences of adult life'.

These statements do of course belong to the general category of 'aims', taken here to refer to the wider ends towards which a school is working. Statements of aims:

- serve an orientation function providing, among other things, guidelines for the planning of curricula;
- involve making value judgements and, as such, will be influenced by external factors, cultural, social, political and economic, which operate differentially from place to place and from time to time;
- represent, therefore, relative and not absolute statements;
- remain as statements of aspiration, at a stage removed from making detailed decisions.

For all their evident lack of precision, the process of thinking about aims should never be omitted from curriculum planning (*see* Chapter 8). Our aims in education are closely associated with the values we hold, and decisions about them are necessarily judgements about what ends we believe to be worthwhile.

The Development of 'Personal Autonomy' as an Educational Aim

Harris (1972) argued that the development of personal autonomy is the overarching educational aim. Of course this is only one of a number of possible standpoints, and would not be acceptable in an ideological setting which did not allow for the right of independent thought and choice as, for example, in a system where the over-riding purposes were to inculcate religious or political truths, and to testify to the infallibility of a religious or political leader.

In singling out autonomy as the key, Harris makes four overlapping points:

1 Autonomy implies **knowledge**, without which responsible choice, a hallmark of an autonomous person, has a fragile base. Knowledge in this instance means far more than acquiring specific facts, and includes such frequently stated intellectual aims as 'promoting the acquisition of thinking skills'; 'forming concepts and principles'; 'fostering scientific methods of enquiry'; 'cultivating discriminatory, critical and imaginative thinking', and so on. Neither should these be seen as narrow academic attributes: rather as contributing to personal development in a wider sense.

2 Autonomy implies the **availability of interests** which an individual wishes to pursue. A person should have an inclination to engage in some valued activity, and a school should aim to provide opportunities for this to develop. A motiveless person cannot be regarded as autonomous.

3 Autonomy is also related to the **satisfaction of needs**. Leaving aside basic physiological requirements, these include:

 – the need to achieve;
 – the need to have interests (a cognitive drive); and
 – the need to belong to or affiliate with (a parent, teacher, friend or peer group).

A dilemma is presented, however, in that satisfaction of personal needs can be held to imply self-regarding ends, which it hardly seems the job of education to promote.

4 The achievement of autonomy must, therefore, presuppose the development of an awareness of and a positive response to the feelings of others, or **empathy**.

The pursuit of autonomy aim is inevitably linked with moral education.

Harris sees a person lacking in autonomy if s(he) cannot make moral judgements, which by definition are prescriptive in so far as the individual is concerned: i.e. a person making such a judgement in principle must allow it to guide her or his actions. An aim of the school must, therefore, be to help to develop the skills and attitudes required in making moral judgements. Such judgements allow the reconciliation of the satisfaction of personal needs aim, and the development of the child as a social being aim.

Another way of looking at aims is to distinguish **intrinsic** from **extrinsic** aims. Intrinsic aims involve the initiation into worthwhile activities, seen as desirable for their own sake. As we shall see, child-centred aims reflect this type. On the other hand, extrinsic aims are associated with external influences, such as the church, the state or society, seeking to further God's purposes or promote the good of the body politic (White, 1982). It is not difficult to justify the view that the 1988 Education Reform Act was more concerned with extrinsic than intrinsic aims. The central focus of intrinsic aims is the good of the pupil, through the promotion of individual content, though not at the expense of the good of others. In this light, the statements of Harris and White are alike in emphasising the importance of moral education, which must permeate all parts of the curriculum.

The moral dimension of schooling can be divided into three elements: moral training, moral understanding and moral reasoning.

Moral training

This is exemplified by old-time, and indeed present-day, religious **instruction**. Basically, the 'good cause' is seen as justifying at its extreme a catechetical type of instruction through inculcation (the process of forcibly impressing something on the mind through frequent repetition, admonition and exhortation) and indoctrination (the associated content – the doctrine or belief-system or good cause being inculcated).

It is not, however, necessarily easy to distinguish education from indoctrination, because both are in differing degrees involved in transmitting social values (Smart, 1973). Many would assume what we transmit in education should be worthwhile, and that we should transmit it in a morally unobjectionable way. But in indoctrination, we must accept that the protagonists see it as worthwhile to transmit the doctrines they support. Opponents and sceptics, however, find both the doctrines, and also the methods of inculcation with which they are being transmitted, morally objectionable.

In reality, the distinctions between education and indoctrination are easily blurred. The problems are specially evident in such areas as recent history, religion and moral education (White, 1967). Indoctrination remains a burning issue, with ministers of state and right-wing think-tanks claiming, for example, that any teaching of recent political history, peace studies and the like, must almost by definition be indoctrinatory. All teachers presumably have political and moral beliefs. If they state these openly, but at the same time draw attention to the existence of counter-beliefs, they might be relieved of the charge of political or other bias, statutorily required by the 1986 Education Act. But is it possible to indoctrinate without meaning to? How can pupils be prevented from identifying unreflectively with their teachers' strongly-held beliefs and values, whether openly stated or as part of the sub-text? Does the compulsory religious assembly inevitably involve indoctrination? Does the position of the school on the spectrum from prioritising a caring environment to promoting intense competition constitute an incipient form of indoctrination, even though overt inculcation strategies are not used? How can it be disproved to the outside world when a class of students from a deprived inner-city school emerges with die-hard views that capitalism is evil, and from a comfortable suburban one that Tory freedom and competition work, that something more than a detached educational values agenda has been offered within? Surely the attempt must be made to apply some basic tests of difference.

Education is associated with a **growth model** which gives primacy to personal development and choice. Indoctrination is associated with a **moulding model**, in which the teacher imposes her or his own opinions and values on the pupil. In striving to avoid indoctrination when dealing with the values dimension, it is important, as will be discussed below, that alternative viewpoints are presented and verification offered. Above all, the educator is watching out for evidence that pupils are beginning to think for themselves about value questions. For the indoctrinator this is the start of the trouble. To the educator, children are the ends in themselves: to the indoctrinator they are means to ends. The essence of education is that beliefs can and should be challenged. In indoctrination the infallibility of the doctrine being inculcated is an essential priority.

There is no doubt that within its self-imposed limitations indoctrination can be very effective: the 'give me a child until he is 7 and I will give you the man' principle. Many in education today would claim that officialdom has too evident a preference for instruction/training rather than true education. The object of the exercise is to produce compliant and passive subjects, workers or worshippers. The ends are manifestly extrinsic, and intended to promote ready obedience to authority: a 'clerkly diligence'. Moral training is in this framework an acceptable means to these desired ends.

Moral understanding

This is an intrinsic aim, to be promoted for its own sake. Among other things, it requires an exploration, both cognitive and affective, of the

values which underlie morality. Fien and Slater (1985) have usefully summarised a number of approaches to work on values, in the context of geographical education.

Values clarification refers to the encouragement of pupils to review their values, to express them openly in discussion, and to respond to challenge from the teacher and peers. The approach has been criticised in that it suggests that values are subjective and all a matter for individual choice: that everybody's opinion is as good as everybody else's. It would further tend to foster the primacy of self-interest, promoting the idea that personal contentment is the end-product. The criticism is less damaging if values clarification is seen as only one strategy in the exploration of values.

Values analysis has the objective of equipping students with the capacity and inclination to make rational and defensible value judgements. A procedure for accomplishing this involves:

    (a) identifying the decisions to be made to resolve the value issue;
    (b) assembling the purported facts on the issue;
    (c) establishing the veracity of the purported facts so that decisions will be based on objective evidence;
    (d) establishing the relevance of the facts and removing distracting information;
    (e) arriving at tentative decisions;
    (f) testing the value principles involved in the decision (Finn and Slater, 1985, p.177).

Further necessary capacities are to be able to:

    (g) comprehend the laws of logic and evidence;
    (h) differentiate fact from opinion;
    (i) distinguish between relevant and irrelevant data;
    (j) test and verify factual claims;
    (k) state explicitly personal value criteria in assessing alternatives;
    (l) test individual consistency in applying evaluative criteria.

It can be argued that in such analysis there is too much intellectualisation of what inevitably are also subjective and emotional issues.

Moral reasoning can be viewed as having similar objectives to values analysis but pursuing them in a less formal, structured way, using group discussion procedures. Case studies are presented in which there are moral dilemmas to be resolved. Kohlberg's theory (1976, p.180) is that there are a series of developmental stages in moral reasoning from selfish childhood egocentrism to a stage of mature moral decision making that is based upon the application of universal ethical principles, such as social justice and general human rights. The hierarchical nature of Kohlberg's 'stages' and the associated pedagogy of group interaction has seriously been questioned, but the basic principles are of interest, not least to geographers exploring social and environmental issues.

Values probing is an approach aimed at integrating the personal valuing, rational analysis and group discussion procedures of the first three approaches. Here the students are expected to:

(a) react to an issue from the perspective of their own values;
(b) clarify and evaluate their own arguments and those preferred by alternative viewpoints;
(c) probe the values which underlie these arguments;
(d) modify their stances or change their minds if so impelled;
(e) as appropriate, act upon their values as maintained, modified, or more radically changed.

The importance of values, which in turn influence personal beliefs and attitudes will be returned to later in discussing the affective domain of geographical education (*see* Chapter 9).

Moral dispositions

These go beyond moral understandings in that they demand more than an armchair theoretical understanding of moral principles, and involve the translation of these rules into the real world as a matter of personal habit: that is, the principles are prescriptive on the person to act upon them.

The way in which they are acted upon will be a matter of degree. Some people are amoral, their disposition being to act purely in their personal interest, unconcerned about the effect this will have on others. Others follow a malign and negative morality which gains satisfaction from the suffering of others. On the positive side, White makes three further distinctions (1982):

• *Minimalist morality* reflects a trimmed-down disposition, in which the moral demands on the individual are kept to a minimum. Such individuals will normally accept other people's basic rights and interests, and follow traditional moral rules – such as caring for themselves, close relatives, and perhaps valued neighbours – but are less responsive to the plight of those beyond their immediate circle, such as, for example, the poor in general, at home and abroad. Many people in western society would be in this category.
• *Universalistic or maximalist morality*, on the other hand, is prescriptive. It may be based on the Christian principle of loving one's neighbour as oneself. If neighbours are defined as all others, this clearly calls for a generous and benevolent moral outlook, likely to engender much personal inconvenience and expense, in a way a minimalist morality does not. It fills the agent's life with obligations (White, 1990),

8

whether pursuing animal rights, famine relief or campaigns against political oppression. It is manifestly in tension with the personal autonomy aim of education, in that autonomy embraces the right to follow self-regarding ends.

- *Concrete morality* is presented by White (1990) as an intermediate form, accepting the unselfishness underlying universalistic morality, while rejecting its unlimitedness. It suggests that charity might begin at home, but should not end there. Each part of our lives is nested in a wider setting: home, locality, community, town, region, nation, and humanity as a whole. But a stress on localising benevolence within smaller scale communities, with which personal well-being might be seen to equate with a broader well-being, is not consistent with universalistic concepts of world citizenship, for example.

While there can be no universal agreement on resolving these tensions, the over-riding argument that moral education should suffuse the curriculum and other parts of schooling can hardly be disputed. This discussion, though inevitably selective, clearly implies an important role for geography in moral education. In the context of the narrow values orientation of both subject and cross-curricular areas of the National Curriculum (*see* Chapter 10), geography is very largely left with the responsibility of globalising the moral and values aspects of the curriculum, thereby moving beyond the limits of minimalist morality.

Educational Traditions: Evolving Criteria for Good Practice

As has already been mentioned in passing, values are closely associated with beliefs and attitudes, and these in turn are formalised in **ideologies**, which are systems of ideas embodying strongly held beliefs and/or traditions. An ideology can have a positive connotation, expressing a benevolent world-view, but more often reflects a distorted vision of reality, actively promoting the vested interests of particular groups. Ideologies form the basis of negative stereotyping of other peoples (*see* Chapter 8). Extremist ideologists characteristically use 'libel by label' techniques to denounce alien counter-ideologies. The more intensely held the ideology, the more rigid the stance taken. The current context of educational practice is manifestly one of warring ideologies. Yet educational ideologies can be framed in constructive terms, so long as they are kept in balance and under critical scrutiny. They embody important fundamental ideas which are the basis of aims statements. They generate different types of curriculum theory and models of good practice.

Educational Ideologies

Skilbeck (1976) has identified four types of educational ideology: classical humanism, progressivism, utilitarianism (bureaucratictechnicist), and reconstructionism:

- *Classical humanism* sees education as the process of handing on the cultural heritage. The general ideology embodies a spectrum of opinion, but generally concentrates on traditional values and is knowledge/understanding-centred. It underpins the subject-based curriculum. In the discussion which follows, subjects like geography are seen as part of the cultural transmission which, at worst, can reflect a narrow élitism and academicism. But, as we shall see, cultural transmission need not be equated with such limitations.
- *Progressivism* is an ideology which is overtly child-centred, based on the goal of enabling the child to discover things for itself and follow its own impulses. Childhood is regarded as important in its own right and not as a preparation for adulthood. In its extreme form progressivism implies that all beings, not least children, are intrinsically good but become corrupted by society. It is in accord with personal autonomy as a basic aim of education.
- *Utilitarianism* is an ideology which looks to utility as a basic end of schooling. An increase in people's well-being is likely to be equated with wealth accumulation and consumerism. In social terms, the central aim is for schools to train pupils to contribute unquestioningly to the nation's perceived economic needs. The utilitarian characteristically attacks progressive ideology on the grounds that over time the educational system has failed to meet these needs and, indeed, has fostered an anti-industrial spirit (*see* Wiener, 1981). Concepts of instruction and training are central to the associated pedagogy.
- *Reconstructionism* is basically a society-centred ideology, though not necessarily one in conflict with progressivism, in that individuals and society are regarded as being capable of being harmoniously integrated. On the other hand, extreme progressive views emphasise forcibly that regarding pupils as future adults is undesirable. A reconstructionist curriculum lays stress on social values and, clearly, on moral education. Subjects are likely to be seen as less important than integrated social studies.

A danger of such categories is that each one may be regarded as mutually exclusive, as labels to which pieces of educational practice, to legitimate them, must be attached. They form the basis of polarisation and stereotyping. In reality, the categories are over-lapping and, in some cases, are as complementary as they are competitive.

Thus it is instructive to weigh in the balance these ideologies as implicit in one of the more extended official statements of aims, that in the Department of Education and Science's *The School Curriculum* (1981):

- to help pupils to develop lively, enquiring minds, the ability to question and argue rationally, and to apply themselves to tasks, and physical skills;

- to help pupils to acquire knowledge and skills relevant to adult life and employment in a fast-changing world;
- to help pupils to use language and numbers effectively;
- to instil respect for religious and moral values, and tolerance of other races, religions and ways of life;
- to help pupils to understand the world in which they live, and the interdependence of individuals, groups and nations;
- to help pupils to appreciate human achievements and aspirations.

These infer some kind of synthesis of the four traditions. As an aims statement it is less attenuated than that previously noted in the Statutory Orders of the National Curriculum.

One problem evident in the past was that particular ideologies became, at various times and in various phases of schooling, over-dominant and out of balance. For example:

- *utilitarianism* was endemic in the old elementary tradition of Victorian times and after, and is clearly an ideology in need of reviving in the eyes of certain government ministers and their right-wing advisers;
- *progressivism* was particularly proselytised during the 1960s and 1970s as the 'one best system', particularly in the primary phase, though it has never achieved the level of prominence here that right-wing opinion has asserted;
- *classical humanism* underpinned the élite grammar school system, as it continues to do in the independent schools and the residual state grammar schools: this is the other strand favoured in right-wing ideology. Here there is a current social twist that takes us back to Victorian times, with utilitarianism seen as the appropriate tradition for less able and especially working class children, its vocationalism contrasted with the classical humanism directed at the more able, and those coming from 'deserving homes', for whom this more academic approach is seen as compatible.
- *reconstructionism* was and remains particularly evident as an ideology associated with comprehensive schooling: it is also favoured among those who in geography and other subjects argue for more radical welfare-based curricula. Like progressivism, this is an ideology which has created high levels of anxiety in the Tory governments of the 1980s and 1990s.

The priorities attached to these traditions profoundly affect classroom practice. They are the basis of aims, and express what their supporters believe to be those most worthwhile. In relation to the three key variables underpinning the structure of this text, classical humanism is most closely associated with the subject variable; tensions between utilitarianism and progressivism with the educational; and reconstructionism with the social.

The contention here is that all these traditions, even utilitarianism, if kept in balance and appropriately applied, can contribute basic principles for good practice, as, for example, associated with the key questions of the curriculum planner long ago identified by Tyler (1949), which lead the process from theoretical aims into classroom practice:

- What educational purposes should the school seek to attain?
- What educational experiences can be provided that are likely to attain these purposes?
- How can these educational experiences be effectively organised?
- How can we determine whether these purposes have been attained?

This procedure has been criticised as being too narrow (*see also* Chapter 5), in stressing outcomes. The text makes it clear, however, that it is also about educational experience. It is vital that these questions are not narrowly interpreted, but are seen in the light of the broader values issues to which this chapter has also been addressed.

Section A

GEOGRAPHY AND GEOGRAPHICAL EDUCATION

CHAPTER 2

The Changing Nature of Geography

The Place of Geography

There has over time been a high level of agreement that geography is about place and space, and the interaction of people with environments. There has been much less agreement over the relative emphasis to be given to each element. For example, place emphasises the particular, and space the general. At times place and space have been seen as complementary; at others, as competitive concepts, with different underlying methodologies. As we shall note, the revolution in the 1960s was largely about the shift from place to space. The case for geography as a distinctive subject has often been based on the fact that, uniquely, it is a spatial subject.

This is a challengeable claim. It can be argued that the case for a distinctive geography is a composite one, based on more than one academic tradition. Space alone is not a concept distinctive to geography. Other disciplines, from atomic physics to astronomy, from geometry to architecture, study space. The obvious difference is the **scale** at which spatial patterns are studied, what Harvey called the 'resolution level' (1969, p.485). Geography's spatial resolution level is a wide but not unlimited one, ranging from the local, through regional, national and continental scales, to the global. Another distinctive quality is the unique interest geography takes in place and space as human territory, focusing on the inter-relationships between particular physical and social environments. Finally, the criterion of mappability is one that is central to the subject. Distinctiveness accrues, therefore, from a whole assemblage of criteria.

The Traditions of Geography

In 1963, Pattison outlined the complex nature of geography on the basis of four continuing, complementary, but overlapping traditions. These were:

1 *The spatial tradition*, which reflects the geographer's concern with spatial patterns, its links with direction and movement between places, and its special attachment to the use of maps as tools;

2 *The area studies tradition*, which reflects the idea that the geographer is centrally concerned with the unique character of places, embodied in the regional approach;

3 *The man–land tradition*, now better expressed as the *people –environment* tradition, concerned with the inter-relationships between human beings and their environments;

4 *The earth science tradition*, which emphasises the description and explanation of the physical features of the earth's surface.

While these definitions may need some modification, they remain useful, not least in avoiding the either–or polarisation in which a new 'ism' is presented triumphantly as a much better system than the previous one. As Stoddart has suggested, notions of revolution and paradigm change are academic means of bolstering 'the heroic self-image of those who see themselves as innovators' (1981, p.78). The grandiose new paradigm becomes a baseline for disciples propagandising the frameworks and the findings of their mentors. There are arguably more fruitful and complex avenues of exploration in the potentially constructive tensions between paradigms. We should, however, first look at the so-called revolution in academic geography of the 1960s, as a dramatic example of a so-called paradigm shift, a paradigm being defined (Kuhn, 1962) as a generally accepted set of assumptions and procedures which serve to define both subject content and methods of scientific enquiry. Like a volcanic eruption, a paradigm shift changes the academic landscape, at least for a time.

The Quantitative Revolution of the 1960s

Idiographic and Nomothetic Approaches

The basis of the paradigm shift underlying the quantitative revolution of the 1960s lay in the switch from idiographic to nomothetic approaches in the study of geographical phenomena. Idiographic refers to the empirical study of unique, non-recurrent events, best exemplified in the area studies tradition. This was devoted to the 'real world' and accounts of different

places and regions. It was concerned with the examination of a wide range
of variables in single areas. The nomothetic approach was scientific in
nature, engaged in the search for patterns and processes, for generalisations
and laws. It was concerned with repeatable events, and involved the study
of a single variable or small number of variables over a wide range of areas.
It thus related more to systematic than regional geography.

One impulse behind the quantitative revolution was a high degree of
disenchantment with what were seen on the one hand as the static
frameworks of traditional regional geography, and on the other the past
excesses of determinism: the idea that the physical environment
determines the nature of the human response. These had lowered the
status of geography to the extent that it was seen more as a school- than
university-level subject. A new approach was urgently sought, and was
found in appeals to natural and social science.

Scientific Methods of Explanation

These are particularly associated with the nomothetic approach. The
object of explanation in science is to establish general laws explaining the
behaviour of events (Fitzgerald, 1974). Two basic approaches are
available: the **inductive**, starting with particular cases and moving
towards universal statements; and the **deductive**, starting with a universal
statement and then making deductions about sets of events. The inductive
approach is regarded by some as the weaker route to explanation. There
are clearly differences between natural/physical and social sciences. In
the social sciences, 'laws' would inevitably be less 'universal' than the
basic laws of physical science. Even so, social science type deductive
approaches grew to be of central importance in the new geography. The
application of statistical techniques to geographical and social science
data became ubiquitous.

The Systems Approach

Another element of this new geography was the enthusiasm for a systems
approach and, particularly, an **ecosystems** approach, seen as peculiarly
appropriate to the subject. A system has been defined as a set of objects,
plus their attributes, plus inter-relationships. The 'set' is in some way
organised by the inter-relationships of the units. In a systems approach the
stress is on functioning, on process rather than on form, on dynamic rather
than static aspects of reality (Langton, 1972). Consequently, anything
which affects the working of the system must be regarded as relevant.

Stoddart viewed the ecosystem concept as bringing together
'environment, man and the plant and animal worlds within a single
framework, within which the interaction between the components could be

analysed' (1965, p.243). The problem in geography was that a system such as 'Britain' is too large and complex to be handled and, therefore, has to be broken down into simpler systems at the level of, for example, a settlement pattern or even a farm. 'The environment' also can be seen as a higher order system of which a lower order system is a part. An example of a lower order system is a farm, which, of course, is part of two higher order systems: the biosphere and the economy. Stoddart further argued that an ecosystems approach was a major step forward, linking geography with the mainstream of scientific thought, providing a basis for inter-disciplinary study, and offering a means of applying theory to real problems. In this advocacy he was supported by one of the architects of the new geography, Haggett, who concluded that '…Geographers are concerned with the structure and interaction of two major systems: the ecological system that links man and his environment, and the spatial system that links one region with another in a complex interchange of flows' (1972, pp.xiv–xv).

The Behavioural Aspects of Geography

From the start, there was resistance to the radical new approaches of the quantitative revolution. Some of this came from those anxious to preserve the traditional values of the regional paradigm. Others regarded the changes as a replacement of one flawed system by another. In particular there was revulsion against the dehumanisation of the subject, scientific and statistical approaches being seen as giving ascendancy to the aggregate and downgrading the individual human agent. The notion that there was a behavioural dimension in geography had already been given some attention, not least by Kirk (1963), who divided the world of geography into two major sub-sections, the 'behavioural' and the 'phenomenal' (Figure 2.1).

GEOGRAPHICAL ENVIRONMENT			
	BEHAVIOURAL ENVIRONMENT	Development of geographical ideas and values	Cultural environs of geographical ideas Socio-economic processes and changing environmental values
		Awareness of environment	Changing knowledge of man's natural environment
	PHENOMENAL ENVIRONMENT	Physical relics of human action	Sequent occupance of environments Man as an agent of environmental change
		Natural phenomena	Organic processes and products (including human populations) Inorganic processes and products

Figure 2.1 The behavioural and phenomenal environments

The concept of the phenomenal environment included therefore not only the natural environment, but also the environment as altered and in some cases made by human beings, in other words, the cultural environment. The concept of the behavioural environment was of course much less familiar to traditional geographers than that of the cultural environment. The important component driving behavioural geography was human agency: the element of decision making. Decision making was seen as responding to personal perceptions of the environment, for people's decisions were influenced not so much by the environment as it was, but by how it was perceived. These perceptions were in the early 1970s linked with the idea of 'mental maps'(Gould and White, 1974) or 'cognitive maps' (Downs and Stea, 1973, 1977). Downs and Stea (1973) explained the process as follows:

> 'Cognitive mapping is a construct which encompasses those cognitive processes which enable people to acquire, code, store, recall and manipulate information about the nature of their spatial environment. This information refers to the attributes and relative locations of people and objects in the environment, and is an essential component in the adaptive process of spatial decision making.' (p.xiv)

Thus on a familiar route to work cognitive maps are used, however subconsciously, as a basis for making the many decisions involved on such a journey. As part of the cognitive map, a series of attitudes, preferences and traits are present. An unfortunate past incident, for example, may colour one's attitude towards a certain route and so another route is, from other points of view illogically, preferred. This preference may harden and become a trait, which is longer lasting than a preference. Cognitive maps are frequently value-laden, and prone to incompleteness, inaccuracy and distortion. Ambrose (1969) identified cultural background, level of education, tastes, value systems and age as significant factors in influencing cognitive or mental maps. These concepts have of course major pedagogic significance, to be considered in Chapter 6. The scope of behavioural geography thus includes:

- environmental perceptions;
- attitudes and responses towards the environment;
- environmental space preferences (for residence, holidays, etc.);
- environmental perception as it affects decision making.

Geography and Social Issues

Interest in systems thinking and the behavioural aspects of geography in the 1970s widened the scope of the subject beyond even its previous broad bounds, with increasing attention paid to social welfare issues. Smith noted 'a shift away from the mechanistic approaches of the quantitative and model-building "revolution" towards more involvement in contemporary social issues along with a renewed interest in applied geography and public policy'. He, and many others to follow, argued that human geography had a social axe to grind, and an applied purpose in helping to improve the human condition (1974, p.289). The characteristic geographic contribution was to appraise spatial distributions from the point of view of social justice. The derived concept was 'spatial injustice', seen as becoming 'one of the major themes of the new socially activist geography', the key questions being – who gets what, where, and how effectively? (pp.295–6) Smith's conclusion was that the 'challenge to contemporary geography would appear to be how to design an education which combines the technical strengths of the quantitative and model-building era with a passionate concern for the condition of mankind' (p.297).

The 1980s and 1990s

One of the striking features of the quantitative revolution of the 1960s was the concern to disseminate its approaches into schools, and these have undoubtedly had some impact since. In the late 1980s and early 1990s, there has been a similar outpouring of geographical texts that could broadly be termed methodological. In 1989 alone, these included Gregory and Walford's *Horizons in Human Geography*, expressly designed to make available to schools a range of new thinking at the frontiers of the subject. There was the geographer David Harvey's *The Condition of Postmodernity* which, among other things, served to place geographical thinking in the broader arena of post-modernist thought. Agnew and Duncan's *The Power of Place* was designed to draw together the geographical and sociological imaginations, and exemplified the greater interest of social scientists in place. Kobayashi and Mackenzie were editors of a volume entitled *Remaking Human Geography*, and Peet and Thrift of *New Models in Geography*, both collecting new thinking related to the competing 'isms' in geography and social science of the 1970s and 1980s. The second edition of Johnston and Taylor's *A World in Crisis? Geographical Perspectives* appeared (1989). In addition, 1989 saw the publication of Bird's *The Changing Worlds of Geography: a Critical Guide to Concepts and Methods*; Cooke's *Localities: the Changing Face of Urban Britain*; Wolch and Dear's *The Power of Geography*; Soja's *Post-modern Geographies: the Reassertion of Space in Critical Social Theory*; and Taylor's *Political Geography: World Economy, Nation-state and Locality*. Notable additions of the 1990s have included Unwin's *The Place of Geography* (1992), Mannion and Bowlby's edited collection on *Environmental Issues in the 1990s* (1992), Livingstone's *The Geographical Tradition* (1992) and Buttimer's *Geography and the Human Spirit* (1993).

It is doubtful whether this outpouring of ideas has as yet had much impact on school geography. Indeed, it does not seem to have had the same impact on geographical educationists as the revolutionary ideas of the late 1960s. To return to Rawling's statement (1993), this necessarily means an impoverishment in curriculum planning in geography. Unlike the case of the late 1960s and early 1970s, when there was curriculum revolution equally exciting to geographical educationists, perceiving a convergence of formative ideas in both geography and education, in the late 1980s it was the narrower horizons of the National Curriculum which preoccupied teachers and those in teacher education. Not much attention has recently been paid to frontiers thinking. Of what relevance could it be when we were being invited to return, it was said, to the dark ages? In fact there was, perhaps unwittingly, one critical aspect common both to the new thinking at the frontiers of the subject, and what emerged from the National Curriculum Geography Working Group: that was a revival of the centrality of place in geography. Gregory and Walford on the one hand stressed the need to come to terms with place, space and landscape 'as never before' (1989, p.3) and on the other the importance of renewing school and university links, school subjects being seen as the poorer without 'that vital, organic linkage' and awareness of the 'excitement and potential of contemporary research' (p.4). Like Geikie over 100 years ago (1882, p. 24) and Olive Garnett over 50 years ago (1940, p. 171), they highlighted the 'linking thread which ties together the different curiosities about the world of the primary school child and the post-graduate researcher' (p.3).

Ideology Recognition

Part of the false consciousness of the quantitative revolution of the 1960s was its confident internal perception that it represented certainty, progress, and neutrality on value issues. It had, it would seem, only a dim perception that it might rather reflect a narrow and ephemeral ideology. One reason may have been that the prevailing climate of thinking referred to traditions (a safer term), as in Pattison's traditions, rather than ideologies. Thus we wrote even in the 1970s of competing regional, quantitative, behavioural and welfare traditions.

There are positive and negative definitions of ideology, as outlined by Baker (1992). Thus on the positive side, he argued that an ideology represents a lattice of ideas which permeate the social order, constituting the collective consciousness of each epoch. On the negative, it could be defined as a false consciousness which fails to grasp the real conditions of human existence. Baker explains the persistence of ideologies as a response to human needs of a quest for order; simplicity rather than complexity; and certainty rather than ambiguity (pp.3–4). They furnish assurance, assert authority, and seek permanence They generally underpin the views of particular interest groups, and their struggles for power (p.5).

They offer a sense of belonging, and (in the real world) enable convenient escape from critical reflection and differentiated reasoning. They are vitally important to the politician, who wishes to preach clear and simple messages to the converted. As we shall see later in the section on social stereotyping (*see* Chapter 8), they are more potent and difficult to dislodge where more extremist beliefs are held. But as Baker has shown, ideologies have an impact not only on the behavioural, but also phenomenal environment as, for example, in the effect of the major ideological change from feudalism to capitalism (p.10).

Johnston (1991 p.31) has offered a simple diagrammatic connection between methodology and political ideology (Figure 2.2).

In the centre of the diagram is the dimly lit "central table" (CT), overlapping the different ideologies, and drawing ideas from them. Thus the HL table reflects various forms of radical, Marxist thinking, and sees knowledge and action as inseparable; whereas the HR is, among other things, logical positivist and neo-conservative, and advances the ideas of western capitalism, though not as an activity for academics. The SL might be seen as a branch of the liberal–humanist ideology, left of centre, and the SR more or less the same thing, but to the right. The soft end is seen as more open to being affected by evidence and rigorous analysis, while the hard end makes itself more audible and visible (pp.31–2).

		POLITICAL IDEOLOGY	
		Left	Right
	Hard	HL	HR
METHODOLOGY			CT
	Soft	SL	SR

Figure 2.2 Methodology and political ideology

Let us now consider some of the key ideologies (rather than traditions) of the 1970s and 1980s, as they affect and are permeated into academic geography.

Humanist Geography

Humanism relates back to the behavioural tradition of geography. Simply, it is more directly linked with historical than scientific approaches. It is closely identified with the concepts of **hermeneutics**, an approach giving priority to understanding rather than explanation. Hermeneutics:

- emphasises the importance of individual human agency, its actions and the meanings underpinning them;
- shows how cultural life is embedded in practical activity;
- concentrates on the explication of contexts rather than striving for universal explanations;
- sees knowledge as a way of coping with rather than of representing reality;
- identifies human judgement as part of the process of interpretation and consequent decision making.
(Based on Atkins, 1988.)

Thus humanism gives more emphasis to personal arenas and subjective influences, to a world of complexity and diversity and, of course, so far as geography is concerned, leads inevitably back to the idea of the central focus of geography as people interacting with people in particular places. To the teacher, especially of younger pupils, the idea that there are geographies of everyday lives, including the lives of children, is of great pedagogic significance. Eyles (1989) sees the key concepts of such geographies as time; space; self; others; interaction; biography; situation; power and structure. Thus we conceive and practice our own lives in time and space, and respond to the actions of others on the basis of social interaction (p.115).

Behavioural and humanistic subjective geographies also connect with the world of landscape, subtly different from the concept of environment. Relph regards landscapes as both forming the contexts of everyday life and also giving meaning. The idea of landscape is seen as vital, subtle, specific and incapable of exact definition. To judge a landscape on standardised appraisal procedures is, in Relph's words 'like judging wines by their alcohol content: accurate but misleading'. The geographical approach to landscape should include the aesthetic and poetic. The skills to be developed in appraising landscapes are:

- ways of seeing (observational skills);
- ways of thinking (through critically reflective insights);
- ways of describing (1989, pp.159–60).

Through such skills we may, as Sauer (1925) long ago postulated, move from seeing the landscape not only as an assorted collection of phenomena, but also in terms of association and connection (in Unwin, 1992, p.98). These skills he saw as potential contributions to a constructive and just use of the environment, and to the better design of places and the cultural landscape. These ideas have interesting implications for cross-curricular links in school, suggesting geography's aesthetic as well as scientific and historical connections (*see* Chapters 4 and 11). They represent the philosophy most compatible with the thinking behind this book (*see* Chapter 14).

There have been many critiques of humanist–hermeneutic approaches. To the scientific mind, they may be regarded as soft and romanticised, and inadequate in explaining the real world (Unwin, 1992, p.158). To the radical geographer, they are seen as having tended to:

- ignore external structural constraints within individually constructed worlds of experience;
- be reflective and passive and lacking in concern for the mechanics of social and political change.

Critical or Radical Geography

Critical geography is in essence part of a wider social science complex which sees itself as going beyond analysis and reflection into social and political action. Its goals are the transformation of society and the emancipation of the individual. It clearly is part of the welfare tradition of geography which emerged (though not for the first time) in the 1970s. But humanist geographers would also see their work as associated with human welfare. The critical ideology goes beyond this and is concerned to achieve a better world through, if necessary, imposed political change.

Radical geography

Radical geography is fuelled by **Marxist** thought and, by definition, antipathy to capitalism. It emerged in America in the period and aftermath of two crises of capitalism: the Vietnam War and urban ghetto riots. Marxist ideas do not, however, form a unified category. In recent decades strong efforts have been made to distance Marxist thought from real-world Stalinist iron-fist applications. Classical Marxist thought is uncongenial to humanist thinking, in being, like logical positivism, about aggregates, macro-explanations and, in this case, deterministic social (and socialist) macro-theory. The focus of fundamental Marxism is historical materialism: the central concepts are modes of production; the struggle of the forces and relations of production as between social classes; the oppression of ordinary people by these modes of production and the associated class structure; and the inevitable change over time to a less differentiated society.

There is general agreement that there has been a significant shift in academic Marxist thinking during the 1980s, to a more sober, less determinist, less certain and less combative form. Both capitalist and socialist modes of production and social organisation are regarded as diverse. The change is explained by Peet and Thrift (1989, p.7) as a result of:

- the powerful critiques of extreme forms of Marxism – as in Gregory's 'brute globalism' (1989, p.v);
- greater knowledge of existing socialist countries making revolutionary politics less certain and less congenial;
- the laid-back academicism of the 1970s being replaced by a narrower professionalism in the 1980s;
- the 'young Turks' of the 1970s becoming part of the establishment of the 1980s.

Among the more recent aspects of Marxist thought is the acceptance of the importance of place. Place and space are seen as expressions of the economic and political system and social structures. As a result there is uneven development, not entirely related to the economic possibilities of a particular area. Some places are privileged by government and the economic system; others not. Class character varies geographically. It is much easier for a government to deal with an alienated proletariat that is spread out than one which is politically rock solid in, for example, its mining region; a part-time, de-unionised work force is manifestly less of a problem than full-time workers in a large, union-organised factory. Such issues emerged powerfully in Harvey's key text of the 1970s, *Social Justice and the City* (1973), in which urbanism was viewed as based on exploitation, and the urban landscape as irretrievably and irreversibly carved out according to the dictates of capitalism.

The global economic situation has spawned another interesting variant of radical thought, which has appeared under the title of **political economy**, a branch of social science (with a strong geographical input) covering global economic processes such as de-industrialisation, the development of informal economies, the rationalisation of labour, capital and the production system, the key significance of upswings and downswings in world economies, and the profound influence on internal regional and community settings of economic decisions taken in distant places.

In the radical interpretation of political economy, capitalism is judged to be incapable of promoting the welfare of the Developing World, with the pillage of nineteenth-century colonialism being repeated in the neo-colonial exploitation of trans-national corporations, the World Bank and the International Monetary fund. In the new world of global capitalism, the central issues are those of debt crisis, famine and the problems of the world's inner cities. There are clearly counter-views from the right, which see capitalism as a pre-condition for solving the economic problems of the developing world countries. Between these the liberal Brandt Commission-type view stresses the notion of inter-hemispherical interdependence. Corbridge is critical of polarisations between North and South, core and periphery, arguing that most economic rivalries affecting the world economy are inter-national rather than inter-hemispherical (1989, p.358).

A further variant of radical geography is that associated with environmental issues, in which protagonists are engaged in pushing the 'green' movement into **eco-socialism**, based on Marxist, anarchist and 'deep ecology' ideologies, aiming for a new red–green political synthesis (Pepper, 1993).

Feminist geography

Feminist geography is also radical in seeking to deconstruct accepted categories and establish new ways of thinking about women and geography (McDowell, 1989, p.147). Feminist geography has never confined itself to a single theoretical perspective, however. There are variants in feminist thought, which are reflected in feminist geography. Of a number of different strands, four are particularly evident:

- *Liberal feminism* – which argues for equal rights for women, within existing social, political and academic structures.
- *Marxist/socialist feminism* – emphasising the gender element of class relations, focusing on aspects of women's experience in the labour market, including the relationship between production and reproduction and, on the broader scale, the inter-relationships between capitalism and patriarchy.
- *Radical feminism* – emphasising the concept of patriarchy, that is the historical domination of men over women as the key form of oppression.
- *Minority feminism* – including black feminism, critical not only of the idea of white patriarchal society triply oppressing black women (on the basis of class, colour and sex), but also of the nature of the white women's movement glossing over differences between white and black women, and also between women in developed and developing world countries (Weiner, 1994, pp. 51–8).

The rise of feminist geography was reflected in the formation in 1980 of a Women and Geography Study Group, affiliated to the Institute of British Geographers, which in 1984 produced a 'state of the art' text, *Geography and Gender: an Introduction to Feminist Geography.* Notwithstanding the presence of theoretical and political variants, basic to feminist thinking is the common theme of the inequality of power between men and women, and the notion that: '...Geography has played a part in the maintenance and particular form of class and racial divisions and patriarchal gender relations are a central part of their construction' (Bowlby *et al.*, 1988, p.6). Challenges to the position were envisaged at three inter-related levels:

- academic, looking beyond the conventional concepts used to explain the geography of society;

- pedagogic, using different source materials and alternate teaching styles for women and girls;
- 'political', publicising the unequal position of women in the educational labour market (1988, pp.5–6).

There are self-evidently common interests as between feminist thought at the academic frontiers and a wide range of issues associated with gender and schooling. Gender issues at school level will be looked at in later chapters.

Post-modernism

Gregory suggests that post-modern is a short-hand term for a hetero-geneous collection of ideas (1989, p.69) that, on the one hand, are a critique of enlightenment values of reason and rationality and, more recently and specifically, reflect the counter-blast of those repelled at the brutalist excesses of Corbusier-type architecture and bureaucratic town planning.

Modernism in town planning was fuelled not by aesthetics but by economics. Its watchwords were efficiency and standardisation, and the primacy of the expert over the ideas of an intellectually under-developed populace. City building became a question of addressing the technical problems of clearance and construction. The landscape which emerged was of high-rise dwellings and offices, and urban motorways. It was a landscape devoid of the critical geographical and aesthetic idea of a 'sense of place'. People were, in this concept, present in capitalist and, on an even more massive scale, in socialist cities, marginalised as aggregate ciphers (Ley, 1989, pp.50–2).

It may be helpful to counterpoint key words which express the contrasting ideas and values of modernism and post-modernism:

Modernism	Post-modernism
Positivist	Heterogeneous
Technocratic	Differentiated
Rationalist	Fragmented
Linear progress	Ephemerality
Absolute truths	Indeterminacy
Continuity	Discontinuity
Stereotyped social orders	Elusiveness
Bureaucratic planning	Nuances
Standardised knowledge	Rejection of theory
Standardised production	Pluralist
(Fordism)	Eclectic
Cubism/grids	Fluidity
Pressure of time	Human agency

The large institution	The smaller unit
Explanation	Interpretation
Obsolescence	Renovation/recycling
Corporate city landscape	The urban village
The important	The everyday
Large political parties	Non-class politics
Central control	(ethnicity, gender,etc.)
Secrecy	Openness
Impersonality	Sense of place

Such listing is obviously dangerous, and suggests the two trends are distinct categories, the latter the property of the 'good guys', and the former of the 'bad guys'. As many have pointed out, there were a host of objections to the values of modernism before the recent post-modernist movement appeared.

The looseness of the ideas collected under the term post-modernism does not, however, mean that it should not be regarded as a significant sea-change in the collective mind, with implications for geographical education. Among the most momentous are technological changes, with their wide-ranging impact on the quality of life of the world's peoples. One of the most disruptive forces is **time–space compression** (Harvey, 1989), made possible by the astonishing technological advances of the last two decades. At the same time old skills become obsolete, as does the machinery that replaces them. Improved and more rapid information flows speed up decision- making, and impact on places and institutions far distant from where the decision was made. The greater speed of flow of goods, services and payments through the market system is made possible by plastic money, electronic banking and computerised trading. There is not only an acceleration in consumption over time, but also across space. The contemporary way of life is associated with disposability (of people, goods and values), volatility, instability (of employment, for example), crisis management, sensory overload (with scores of TV channels and a bombardment of advertising on screen, in the press and through the post), trivialisation and manipulation of opinion (the tabloid press), frenzied life-styles, psychological stress at home and at work, a litigious, buck-passing culture, and the over-riding importance of the image, now transmittable simultaneously over the globe, leading to an internationalisation of values and attitudes: the global village come true.

Many of these changes have little respect for territories, particularly in the context of political economy. The internationalisation of commerce and culture shows little respect for world recession, global environmental problems, and indigenous peoples. Space is regarded as a critical element in this, for as industry moves out of traditional regions, new skills need rapidly to be acquired. In the global system, cities become, or are at least perceived as, being privileged locations, attracting mass immigration. There is a greater tendency for places to be differentiated competitively, some to be made attractive to inward investment of capital, and some not. Of overwhelming importance for the parts which make up the global economy are the instant or rapid decisions related to currency fluctuations, the massaging of vast accounts, the impact of changing interest rates, national devaluations, and the workings of the key stock exchanges, as the financial community lives by the minute and trades by computer.

The internationalisation of consumption is another critical characteristic of post-modern global society. Cross-border migration, the computerisation, as already noted, of commodity markets, and the impact of the media, have produced common culinary styles, pizzas and pastas now being as available in British as Italian cities, for example, while cross-hemisphere as well-as cross-ocean movement of wine has revolutionised production, consumption and marketing of this product. There will come the day when reality is one big screen, and many of the functions of economic and social life can be conducted from a keyboard.

Post-modernism conceives of different place identities, in some of which depressed working classes, racial minorities, colonised people, and women are congregated. In some cases these groups have power within the community, but are place-bound and disempowered outside it. Those living in inner cities pay the highest social costs, not least insurance costs. Advantaged groups are the powerful who move into and control favoured places, their power increased by time–space compression. Some are so powerful that not only small groups, but also large nations, are being disempowered.

As a thought system, Harvey (1989) sees advantages and disadvantages in post-modernism. It is favourable in addressing questions of and in its concern for difference; complexity; nuances of interests and values; the many forms of otherness; spatial diversity. The disadvantages are that it at the same time does not have anything necessarily to offer to assuage fragmentation of social landscapes, the increasing isolation of poverty, ghetto cultures and the realities of the global economy and the associated power structures. As Harvey points out, brutalist architectural forms in the 1960s were not themselves responsible for inequalities in social life (1989, p.115).

It is important therefore that in recognising the significance of post-modernist thought for geography and education, the advantages and disadvantages of previous thought systems are recognised, and polarisation and determinism do not take over. New paradigms that question such values as 'enlightenment' and 'rationality' must surely not be unchallenged in educational discourse. Such reservations are important in discussing the impact or otherwise of post-modernism on geographical education, points that will be returned to in later chapters.

The Binding Logic of Place

Like the issues-based approaches which burgeoned in the 1970s and 1980s in school geography, the various theoretical 'isms' at the frontiers arguably also led away from distinctiveness. To some this did not matter, but in school terms at least a place in whatever curriculum needs to be justified by what a discipline has to offer as an intellectual and social resource. One thing radical, humanist, and post-modernist geography have converged upon is recognition of the centrality of place and space.

Agnew and Duncan (1989) have noted the richness of meaning of the term 'place' as:

- a portion of space where people dwell together;
- a rank in a list, e.g. first place;
- a temporal ordering, it took place;
- position in a social order – knowing your place (p.1).

They look for a concept of place which synthesises both geographical and sociological imaginations, place being the arena of mediation of the lives of individuals and institutions in a range of global societies. In practice, they seek a new regional geography which is much more than an inventory of characteristics, set against an immutable physical backcloth. They require a concept in which spatially bounded arenas of economic and political power are mediated through social interactions specific to particular localities.

It may seem that these frontier developments and Natural Curriculum definitions make strange bed-fellows. But as Daniels has pointed out, the 'National Curriculum refocuses on place' (1992, p.310), while at the same time the study of places is being revived in many different aspects of geographical research.

CHAPTER 3

Continuity and Change in Geographical Education

From the starting point of the established traditions of geography, this chapter focuses on continuity and change in geographical education, and the way in which ideas and materials from the frontiers of geography have been applied over time. One persistent tension has been that between regional and systematic geography, for example. While systematic elements, such as earth science, have perennially been seen as important in the development of geography, it has equally been pointed out that these can splinter the discipline and lead practitioners into the fringe zones and even beyond the disciplinary boundaries. It has rather been the place (area studies) and space traditions that have established themselves as the elements most distinctive of geography. Any systematic study that is distinctively geographical must by definition be tied to real places or to distributions and patterns over space at the different scales which geographers treat. Let us explore these tensions through a brief résumé of the development of geography as a school subject.

Studying Places: 'Hard-core Traditionalism'

The 'Capes and Bays' Tradition

The so-called 'capes and bays' geography blossomed in a period when less and less in the map of the world was *terra incognita*. In the nineteenth century European nations were completing their colonisation of places hitherto unknown to the western world. It was therefore regarded as an educationally valuable activity to learn the names of places, recognise where places were and, moreover, where the places ruled by Britain were.

   At the same time, the teaching of locational knowledge was irrevocably tied to a pedagogy derived from religious instruction, namely catechetical teaching. This combination of locational knowledge as content and rote memorisation as the process of acquiring it is referred to here as 'hard-core traditionalism'. It was distinguishable by its:

- heavy stress on rote learning and factual recall;
- use of geographical facts and factors as ends in themselves;
- reliance on deterministic explanations;
- pursuit of world coverage, attained by superficial listing of geographical information;
- treatment of value-laden information on, for example, national characteristics, as uncontentious fact;
- use of such information to glorify Britain's imperial position;
- heavy reliance, whether at elementary or secondary levels, on external examinations as the main motivating force, reinforcing the work ethic and the competitive spirit;
- clear-cut demarcation between teacher and taught, with lessons dominated by chalk, talk and note-taking, and use of a very limited range of materials, consisting of blackboard, atlases, and catechetical textbooks.

The pursuit of locational knowledge as an aim is today in disrepute, a situation which can in part be traced back to revulsion against the 'capes and bays' image of the subject. It is, however, important to recognise that in this catechetical period the facts were ends in themselves and not merely means to ends. Additionally, the essence of the 'capes and bays' approach lay in the rote learning of these facts. To reveal the true nature of 'capes and bays' geography, the Rev. Goldsmith's introduction to a geographical textbook of 1823 is quoted:

'The proper mode of using this little book to advantage, will, it is apprehended, be to direct the pupil to commit the whole of the facts to memory, at the rate of one, two, or three a day, according to his age and capacity, taking care, at the end of each section, to make him repeat the whole of what he has before learned.' (pp.3–4, quoted in Vaughan, 1972, p.131)

The academic consequences of geography's association with such procedures were damaging to the subject's status at secondary and still more at tertiary level. Political geography, associated even in higher education with an encyclopaedic approach to world knowledge, was of much lower status than the developing earth science branch of geography, that is, physical geography. The Royal Geographical Society publications of the late nineteenth century made clear the difficulty experienced in getting geography established in the great public schools and ancient universities. The Society solicited the views of headmasters on the value of geography as a subject in schools, included in the famous Keltie Report (1886). Many were dismissive, drawing attention to its lack of intellectual rigour as it was then practised:

'The subject is merely an effort of memory. We cannot make a discipline of it, nor set problems in it.... As a compulsory subject in public schools, and taught – as it must be under those conditions – by a number of assistants who have no special interest in the work, and cannot clothe the dry bones, it must fail'. (p.92)

The famous nineteenth-century geographer, Freshfield (1886), described the process more light-heartedly, in reflecting on examining procedures during his own schooling: 'In my day there were many boys...who acted consistently and not altogether unsuccessfully, on the principle that whatever was not a city in Asia Minor was an island in the Aegean Sea' (p.702).

Regional Geography

Two of the great names in the history of the subject, Mackinder and his disciple Herbertson, were well aware of the low estate of geography in academic circles in the late nineteenth century. Mackinder in 1887 spoke of the educational battle being fought to make geography a discipline instead of a mere body of information. Herbertson was still recording (1896) the opposition to the progress of geography teaching in schools, in part a result of its perceived vagueness, and in part its reputation as an *omnium gatherum* subject.

Mackinder and Herbertson were anxious to systematise geography and rescue it from its encyclopaedic image. They deplored the nineteenth-century tendency to divide it into distinct compartments, physical and political. In many schools concentration on political geography placed a burden on the child's memory through the use of a multitude of divisions, such as countries and counties, as units of study. Herbertson (1905) evolved a new and more sophisticated 'model', based on the so-called 'higher units'. In place of a framework of numerous political divisions, he proposed the use of 14 'natural regions'. In pursuit of his duty to be 'not merely descriptive but scientific', the geographer needed to study 'distributions of various forms and forces on the Earth's surface and their inter-relation' (1913; reprinted 1965, p.335). Herbertson lost no time in diffusing his ideas into the schools. His *Senior Geography*, first published in 1907 and remaining in print until 1952, opened with a chapter on the natural regions of the world.

Later textbook writers such as Newbigin, Unstead, Stamp, Stembridge and Pickles were all influenced by Herbertson. Hence in Pickles' *The World* (1939) it is claimed that:

'...it is possible by ignoring minor differences to divide the world into a dozen or so climatic regions.... By this means a survey of the world is made relatively easy since when we have learnt the geography of one region we can easily learn the geography of another region of the same type.' (p.35)

The possibilities for higher-level generalisation provided by this scheme appeared an important step forward. They had this potential. Unfortunately:

• the scheme was implicitly determinist, suggesting that climatic factors in particular determined human responses which, as we shall find, had serious consequences in terms of the negative stereotyping of certain human groups;
• the theoretical attractions of the scheme seem never to have been firmly translated in any widespread way into good educational practice.

By the post-1945 period 'natural regions' geography was barely distinguishable from the 'capes and bays' geography, in the sense that the presentation of material was equally inert and cumulative, and the learning procedures similarly concentrated on memorisation and recall.

Though it may have been no intention of Herbertson's, determinism remained. The 'factors', to be considered truly geographical, had to be largely physical factors or, to some extent economic. Thus the first major conceptual revolution in school geography, the innovative regional geography paradigm, when translated into secondary schools, came to blend too easily with the pre-existing rote learning modes. Geographical factors were added to geographical facts as ends in themselves, effectively devalued as material of legitimate concern in a progressive education.

Studying Places: 'Enlightened Traditionalism'

The Field Work Tradition

It may be a surprise that the field work tradition in geographical education goes back at least as far as the period of 'capes and bays' geography, though its influence has too easily been submerged by stronger forces. Field work was an important part of the nineteenth-century *heimatskunde* (home studies) approach in Germany. By the late nineteenth century, the process of using the environment for educational purposes had a broad backing of educational thought behind it.

The main impetus to field work was through nature study, rural life and grand scenery. An anticipation of ventures to come was provided by a Hampshire teacher, John Dawes, who saw the purpose of education 'to make the children observant and reflective; to make them think and reason about the objects around them.' They were therefore encouraged to look and think:

'They watched the animals in the fields and collected statistics as to how they moved. They followed the progress of the seasons by observing the sun.... Geography was taught in relation to the locality and to the observations of the sky and the weather made by children in their scientific work.' (Ball, 1964, pp.62–4)

It was Sir Archibald Geikie, above all among late nineteenth-century academics, who rebelled against geography, in school or elsewhere, being seen as a mere exercise of memory. Though a professional geologist, his methodological text *The Teaching of Geography* (1887) remains one of the most distinguished books of its type ever written:

'In dealing with the young we should try to feel ourselves young again.... To cement the bond between teacher and taught there should (in the early stages) be no set tasks for some considerable time. The lessons ought rather to be pleasant conversations about familiar things. The pupils should be asked questions such as they can readily answer, and the answering of which causes them to reflect and gives them confidence in themselves.... The objects in the schoolroom, in the playground, on the road to school, should be made use of as subjects for such questionings.... A fact discovered by the child for himself through his own direct observation becomes a part of his being, and is infinitely more to him than the same fact learnt from hearsay or acquired from a lesson-book. The idea of discovery should be encouraged in every way among children.... There is happily now a growing recognition of the principle that adequate geographical conceptions are best gained by observations made at the home locality. The school and its surroundings form the natural basis from which all geographical acquirement proceeds...' (pp.7–11)

Geikie's first fossil collecting expedition to an Edinburgh quarry convinced him of 'the enormous advantage which a boy or girl may derive from any pursuit which stimulates the imagination' (1882, p.24). From local study, the youthful experience needed to be widened, moving out to the national level, and thence 'to broad and intelligent views of the world at large' (p.4). Field work was also regarded by Geikie as an important device linking subjects in the curriculum. Huxley (1877) agreed, and wrote an influential educational text on the subject of physiography, an optional subject in the elementary codes in the late nineteenth century, to which he attached the sub-title, 'an Introduction to the Study of Nature'. Cowham (1900) similarly viewed field work as a means of amalgamating geography, physiography and nature study.

The migration of famous physical geographers and earth scientists into the educational sphere offered to field work both a stronger scientific thrust and greater educational rigour. Davis (1902), discussing its importance at secondary level, for example, stressed that there was more to be done than mere observation. It should be quite clearly understood that work of this kind (i.e. field work) is:

'not limited merely to matters of observation and record. These are truly the first steps, but they must be followed by abundant thinking; indeed, the soul of the work is gained only when the thinking that is inspired by the observations is logical, searching, critical.' (p.329)

There was considerable discussion of the value of education in the open-air in early issues of *The Teachers' Times, The Practical Teacher* and *The Geographical Teacher*. Thus Lucy Reynolds, a teacher at a private school in Kendal, in 1906 vividly outlined the excitements of taking children on a residential visit into the Lake District, 'enlisting the natural interests and enthusiasms of the pupils...and yoking their spontaneous activity and love of seeing and doing as powerful motive forces...' (pp.152–3).

A factor inhibiting the growth of field work was that it was perceived of as essentially a rural pursuit. Urban children were seen as in need of an experience of a world that had been lost for the mass of the people, now subject to the physical pollution and moral corruption of large towns and cities. This increased the problems of conducting field work. The expense of transporting children into the country was another issue. As a Cardiff geography mistress, Joan Reynolds, complained (1901):

> '...money for tram or bus fare is almost a necessity in order to get into the country.... Distance from the country involves not only money, but also time and worry. That great question of our towns comes to the front so often when we attempt educational reforms.' (p.33)

Much earlier, however, Geikie (1887) had argued that while the countryside was obviously advantageous for field work, the skilful teacher should be able to find topics of interest even in a wilderness of streets and houses. The Board of Education (1905) urged the study of home surroundings whether in town and country, as did the Royal Geographical Society, in a series of 'Syllabuses of Instruction' for elementary and secondary schools, reprinted in *The Teachers' Times* (1903). Penstone's *Town Study* (1910) was a pioneering text in this area, yet there continued to be resistance in the inter-war period. Thus even Hadow (Board of Education, 1931), while advocating field work both in town and country, still saw the country child as having an advantage over the town child in having ready access to field study – in the development of mapping skills, for example. Cons and Fletcher provided a breakthrough in 1938, demonstrating that what was possible in the countryside was equally applicable to the inner suburbs. Similarly Coulthard in Bishop Auckland (1938, 1946) sought to promote, through work out of doors, not only geographical skills but also community involvement.

Though the advocacy of field work is therefore a long-standing tradition, diffusion of the idea through the school system was slow to come. Why did teachers at both primary and secondary stages prevaricate for so long a time in the face of so powerfully argued and seemingly irrefutable a case? Apart from the matter of expenditure of time and money, one reason may have been lack of confidence. Nor were all teachers and others agreed on the desirability or effectiveness of field work. Professor Lyde (1912) opposed it on the grounds that teachers could not afford the time, the boys 'regarded it as a picnic...the road to knowledge must not appear too easy' (pp.200–1). The major factor at secondary level, however, was almost certainly its lack of attachment to the external examinations system, where for so long the regional, textbook tradition prevailed, a situation remedied with the introduction of the Certificate of Secondary Education in the 1960s.

From Type Studies to Locality Studies

Perhaps the only significant problem of giving field work primacy is that the local area can become an end in itself. Olive Garnett (1940) rejected parochialism. As she put it, while geography should start with local studies, children must develop their geographical skills by studying 'maps, places, peoples and things beyond their immediate experience' (p.171). Similarly, MacMunn (1926) quoted a 10-year-old child seeing an entitlement denied in the concentrating on the home neighbourhood, which 'kept you waiting to find out all about the world'.

An early means of starting with the familiar, but quickly leading out from it, was to begin with products from different parts of the world. The American textbook writer Miss Willard in 1826 began with the products found in the country store. In a dialogue with his mother, the young Frank asked her to 'fulfil her promise of teaching him geography', desiring to know where the places from which day-to-day products the family used were obtained, where they were and what they were like. 'The cloth for my new coat, my father says, was made by people who live in England' (p.10).

'Object lessons' were similarly designed to introduce the real into teaching about the home country and the world. Unfortunately, in the mass elementary class-teaching situation, the approach often became another rote activity, as Fitch (1884) described: '...one is doomed to hear one subject after another treated in exactly the same way, and to see it solemnly recorded on a board that a cow is graminiverous, or that an orange is opaque' (pp.364–5). Reverting to Miss Willard's approach, Elsa and Dudley Stamp in their *New Era* series of the 1930s stimulated children's interests by starting with familiar objects such as boxes of matches and the ingredients of Christmas puddings, before moving children out into the wider world. A similar idea was followed in Finch's *Geography through the Shop Window* (1931), offered as a continuation of the 'Stories of other Lands' procedure (p.5), officially recommended by the Board of Education.

The idea that geography can bring the world into the classroom through detailed case studies can be found as early as the 1890s in the writings of the American Herbartarians. The German philosopher and educational pioneer Johann Friedrich Herbart argued, among many other things, that the unity of knowledge was of key importance in education and would be promoted by the co-relation of subjects. This could be achieved through, for example, local study, linking science, history and geography. Charles McMurry (1899), a prominent American Herbartarian, drew on and extended this idea in his advocacy of **type-studies**, which:

> '...must be capable of graphic, picturesque treatment. They should be rich in instructive and interesting particulars, not abstract, formal and barren. Our type studies, therefore, must combine two great merits: they must involve representative ideas of wide-ranging meaning in geography, and they must, at the same time, be concrete, attractive, and realistic.' (p.122)

The studies would further have an important comparative function, being drawn from different continents, seen as 'the best possible stimulus for thought, reasoning, and what we might call self-active effort'. While the scale of the places used was not as restricted as later concepts of case-studies or locality studies, each type-study was by definition centred on 'the natural stronghold' of a key geographical idea. They included scales from, for example, a Californian gold mine to the Andes mountains (pp.124–5).

The Herbertsons seem to have been working on similar lines in suggesting in *Man and his Work* (1899) that it was important to start with small-scale studies of simple societies with clear-cut relationships with the environment (preface). Many examples of detailed case studies are present in the twentieth-century textbook literature. One of the best known is the primary series of Archer and Thomas. They sought to take children beyond what they saw as dangerous provincialism and, from a young age, stimulate their interest through intimate studies of the lives of children in distant localities. From every point of view but one the series was a triumph. Unfortunately, Archer and Thomas, in focusing down on 'interest', emphasised the different and the exotic. Similarly, in aiming for clarity, they concentrated also on what were at the time generally accepted as the simpler relationships and lesser needs of more backward peoples (*see* Thomas, 1937), a theme to be discussed further in Chapter 8. At secondary level, similar approaches appeared in the Fairgrieve and Young *Real Geography* series, first published in 1939, and successfully extended after World War II, in which each chapter dealt with a particular place, with generalisations made only after it had been studied.

More and more books followed this trend in the 1950s with again detailed locality studies and broader accounts further afield, as in Forsaith's *Many People in Many Lands* (1951), which included detailed investigation of a colliery district in Fife at home, and the larger scale of Lapland and the way Lapps live, abroad. The Honeybone series *Geography for Schools* (from 1956) offered a ground-breaking example of the systematic translation into textbook form of the sample study method, formally discussed as a methodology in articles by Hickman in 1950 and Roberson and Long in 1956. From that point a flood of such materials appeared in a whole range of texts, many of them very familiar to contemporary geography teachers. The term 'sample study' was later replaced by 'case study' because the idea of a 'sample' was likely to be confused with its more technical use in statistics. But the Geographical Association was still using the term in its booklets *Sample Studies* (1962), and *Asian Sample Studies* (1968). These included farm, village, small-town and port studies. Their central rationale was to support the trend to greater reality in the teaching of geography in both primary and secondary schools. They were essentially locality studies, no new invention of the framers of the National Curriculum.

Such detailed studies have the potential advantage of providing vivid, thorough and accurate detail, in a sense recreating field work in the classroom. Their use is flexible, and their wide availability now makes them a viable prospect for almost any syllabus at any level. One benefit not usually put forward is that they ideally require some original research, unlike the recycling that goes on to make many general geographical materials for schools: what was once termed a 'thrice-boiled essence' of the subject. They also subscribe to Honeybone's (1954) plea to restore balance in geographical education, involving the renewed linkage of human and physical elements of the subject disturbed, as he saw it, in a tendency to over-emphasise physical geography for its own sake. Such linkage could be achieved through detailed place studies (Honeybone, 1954).

If improperly used, such case studies could of course lapse into an analogue of the descriptive world coverage approach, full of trivial facts, signifying little or nothing. They could also offer a distorted view if taken as typical of a particular country, a new form of stereotyping in fact (Chapter 8). Like any other method, to justify their educational existence, the studies had to contribute to the attainment of wider aims and objectives. But in general advantages outweighed disadvantages. They constituted a creative response to Mackinder's 1943 plea to abandon 'natural regions', and to move outwards from 'focal points from which the visualising and rationalising eye can sweep over gradually widening areas' (p.70). Fitting well into both concentric and concept-based schemes, they continue in principle to be a very effective teaching device, as the National Curriculum has, fortunately, recognised (*see* Chapter 11).

The Changing 'State of the Art' in the 1960s and 1970s

The 1960s: the Conceptual Revolution

As we have seen in Chapter 2, the 1960s witnessed a transformation in the 'state of the art' of geography: or rather, it might be said, 'the state of the science'. By the 1960s, it can safely be claimed that 'capes and bays' teaching had very largely disappeared, but that factual recall methods of teaching regional geography were still widespread. In many schools such 'hard-core traditionalism' sat side by side with 'enlightened', in that stereo-typed regional approaches and memorisation of dictated notes procedures were used to see pupils through 'O' level examinations, while the 'enlightened traditionalism' of case study approaches was practised in the lower forms of the secondary school. In some schools, the new quantitative geography was tacked on in the sixth forms, where it was felt to be helpful for pupils wishing to go on to study geography at university. Thus the two identified traditionalist approaches, and the new geography were not infrequently practised in the one school during this transition period.

One of the advantageous features of the 'new geography' was the stimulus it gave to communication between teachers in universities, colleges, and independent and state schools. University geographers assisted dissemination in schools in a way that had not been witnessed for half a century. Joint conferences of school teachers and academic geographers were arranged. The January 1969 edition of the journal *Geography* was devoted to these new developments, and included contributions by academics such as Chorley and Gregory, and also from teachers and college lecturers, who were important influences in disseminating the new geography into schools. They included Everson, Fitzgerald, and Walford. Another important influence was the American High School Geography project, led by a notable American geographer, Helburn (1968), whose materials demonstrated ways of translating the ideas drawn from the academic frontiers fairly directly into school practice. A 'new model army' of textbook writers appeared (Walford, 1989), including Everson and Fitzgerald (1969), Briggs (1972), Fitzgerald (1974), and Bradford and Kent (1977). There was even a pioneering primary series, *New Ways in Geography*, by Cole and Beynon, which is still in print.

There were of course counter-currents running against this tide of enthusiasm for quantification. They included the die-hard traditionalists, whose preoccupation had been with writing massive regional texts. But they also included those who, like Alice Garnett (1969), foresaw a fragmentation of the discipline, and loss of distinctiveness, through the increasing and 'well-blinkered' specialisation she discerned among younger academics.

As with the regional paradigm earlier in the century, so far as schools were concerned, the quantitative revolution was seen as more appropriate to advanced and academic high-flying pupils than those lower down in the school and/or of lesser ability. As already noted, the new approaches involved a switch from inductive to deductive methods of reasoning, from data collection to data manipulation using appropriate statistical techniques, and from unique events to general patterns. Some of these trends (such as the use of deductive reasoning and an abstract methodology of generalisation and theorisation) seemed, like the new regional geography of many decades before, contrary to approved pedagogic practice of moving from the real to the abstract, the simple to the complex, the familiar to the unfamiliar and, especially, the particular to the general. Scarfe (1969) asserted that conclusions must be based on exact down-to-earth evidence. That is why 'the present emphasis on abstract models is pedagogically quite unsound in school' (p.20). Thomas (1970) similarly considered that school geography could not be expected to commit itself to the search for universally valid generalisations, unless it was to abandon its traditional function of introducing the child to the significant variations 'between the several parts of the earth' (in Bale *et al.*, 1973, p.73).

If what was being advocated was a mere dilution of more advanced techniques for school use, then such arguments might have been justified. They were put forward at a time of reaction to some of the more extreme claims being made for the 'new geography'. Similarly, a minority of teachers attached themselves over-enthusiastically to the new paradigm seeing, following their counterparts in the universities, a scientific basis as giving more respectability to geography. The new thinking matured, however, and became less aggressive and more balanced. Potentially, there was the prospect of a fruitful blend for school geography in the infusion not only of mathematical and scientific thinking, but also of concepts from the social and behavioural sciences, which had also adopted quantitative techniques.

Behavioural Geography in School

In the early stages, there were few perception studies applied to classroom use. Preliminary examples included those in Cole and Beynon (1968) and Walford (1973). The Bristol Schools Council Geography 14–18 project designed materials to throw light on people's differential 'frames of reference' (Hickman *et al.*, 1973), that were deliberately ambivalent and subjective, posing problems that invited value judgements, but at the same time demanded that the subjective basis of the judgements made should be explicit and open to discussion. Gould and White (1974) illustrated the potentiality of using short classroom questionnaires for exploring children's 'images of Britain'. A convincing case was made for reproducing such studies for classroom use in order to promote intellectual understanding of environments and peoples and social awareness. Our decisions and attitudes were held to be based on personal 'perceptual filters' (p.45) seen as entrenched early in life and often productive of distortion.

Linking Geography and Education

University Department of Education Syntheses

The influence of developments in curriculum theory on geographical education will be more fully investigated in Chapter 5. Suffice it to mention here that concurrent with quantification in geography, there was revolution in curriculum thinking in Britain. This advocated more rational and structured approaches to curriculum planning, exciting geographical and other educationists, who saw a natural convergence in developments in geography and education, and sought to apply these to planning schemes of work in schools. A number of complementary publications from geographical educationists in departments of education ensued.

During the 1960s, the theoretical basis of 'enlightened traditionalism' was reflected in Long and Roberson's methodological text of 1966, *Teaching Geography*. This not least laid stress on reality in geography teaching, through field work and sample studies. During the 1970s, a shift of emphasis can be identified between Bailey's methodological text, similarly entitled *Teaching Geography* (1974), which emphasised the geographical dimension in geographical education, including the 'new geography'; and those of Graves, who was one of the first to draw together the positive aspects of 'enlightened traditionalism', the 'new geography' and, above all, curriculum theory, in methodological texts with titles subtly different from those of Long and Roberson, and Bailey, such as *Geography in Education* (1975), and *Curriculum Planning in Geography* (1979).

A similar approach was offered in Marsden's *Evaluating the Geography Curriculum* (1976a). Here the structure was based on current models of curriculum planning, highlighting the importance of curriculum theory as well as of geographical content. Like Graves' *Geography in Education*, it explored historical as well as contemporary contexts and, indeed, permeated ideas from the different disciplines of education into an applied form. It also paid more attention than before to the importance of good assessment practice, another educational issue of moment in the 1970s. Hall in *Geography and the Geography Teacher* (1976), covered many similar aspects of curriculum study as applied to geography, and paid particular attention to the examination system and the ongoing Schools Council projects in the subject, as did Boden (1976), who also introduced a school-based element by including articles by practising teachers.

Geographical educationists were also involved during the 1970s in editing a growing series of texts which brought together a range of issues relating to geography and education, as seen through the eyes of different writers. Examples of such texts were the collections by Graves (1972), Bale *et al.* (1973), and Williams (1976). Papers of the first Charney Manor Conference, a geographical gathering of academics, educationists and teachers, were published in Walford's *New Directions in Geography Teaching* (1973).

Schools Council Projects

These methodological texts may not have had the direct impact on geography in schools as was to be the case with the government-funded

Schools Council projects of the 1970s, but were all part of the same period of optimism and innovation. Geography was a particular beneficiary of Schools Council sponsorship, with no less than four development projects. All the projects were led from institutions of higher education:

1. *Geography for the Young School Leaver* – from 1970 at Avery Hill College of Education, London, directed by Rex Beddis, designed to meet the needs of less able children, aged 14 to 16.
2. *Geography 14–18* – from 1970 at the University of Bristol Department of Education, directed by Gladys Hickman, producing major publications in 1973 and 1977.
3. *History, Geography and Social Science 8–13* – from 1971 at the University of Liverpool Department of Education, directed by Alan Blyth, which published its manifesto in 1976.
4. *Geography 16–19* – from 1976 at the University of London Institute of Education, directed by Michael Naish, the Bristol project having in the event concentrated on the 14–16 age range (1987).

The most influential of all these was probably the Avery Hill project, its successes celebrated in a follow-up literature on dissemination procedures (MacDonald and Walker, 1976; Parsons, 1987; Boardman, 1988). The project avoided the mistakes of earlier 'top-down' Schools Council projects and sensitively connected up with local authority networks. It was in a sellers' market too, for there was precious little quality material for less able children at a time when the school-leaving age had recently been raised to 16. There was considerable under-expectation of the less able, perceived as less needy of lavish resources and more up-to-date ideas. The Avery Hill materials were attractively presented, and were based on themes of a high degree of social relevance, thought to be likely to win the interest of young school leavers. Like the Bristol and London projects, Avery Hill saw the need both to promote school-based innovation, yet at the same time to engage with the public examination system, seen as essential if any development for 14 year-olds and above was to take off. As a result, special syllabuses related to these projects were introduced by different public examining boards.

As Chapter 2 demonstrated, the quantitative revolution generated opposition on the grounds of its detachment from human issues and social realities, and not least in schools. The methodological texts and Schools Council projects above in general promoted a more balanced, eclectic and pluralist approach, making use of the benefits of each paradigm. This can be exemplified in the Bristol project's agenda of priorities for classroom change:

1. Greater consciousness of the role of concepts and models in both

geography and everyday thought, pointing up their utility and limitations.

2. Deeper analysis of the processes underlying geographical patterns, especially through simulations, studies of the perceptions and values of decision-makers, and the more sensitive analysis of aggregated data.

3. The use of systems perspectives to emphasise multiple and cumulative causation of geographical patterns.

4. The analysis of 'before and after' data to bring out the key role of identifying basic processes of change in providing satisfying explanation in geography.

5. The more thoughtful examination of practical issues of environmental management, paying due regard to their long-term and political contexts.

6. More study in depth through course work and individual studies.

7. Greater use of resource-oriented and enquiry-oriented styles of teaching (Hickman *et al.*, 1973, p.12).

These far-reaching projects were operating in a transition period in which, as we have seen, there was an increasing scepticism as to the value of quantification in geography, and a growing interest in welfare approaches (*see* Chapter 2). The developments of the late 1970s were thus shifting geographical education towards its preoccupation of the 1980s, the social education function of geography, the subject of Section C of this book. They were also working in a radically changing political context, to be covered in Section D. In the meantime, we need to consider another long-standing aspect of geographical education: its connection with other areas of the curriculum, and its place in integrated approaches.

CHAPTER 4

Perspectives on Geography and Curriculum Integration

Subject versus child-centred?

'Recent work in the social history of education...has revealed very clearly that some subjects have a place in the curriculum primarily because people who have a vested interest in the study and teaching of them...have worked very hard, at a political level, to establish and maintain them there...by such devices as the setting up of subject associations, the drawing up of national syllabuses and, in particular, the establishment of public examinations...a subject such as geography has brought itself...to the point where its right to inclusion in most people's list of core curriculum subjects is likely to go largely unchallenged...we look in vain for an educational justification, or any real discussion of the contribution of the subject to the processes of educational development...the concern is with the knowledge to be transmitted and thus with the acquisition of it by the pupils...' (Kelly, 1986, pp.118–9)

The thrust of this critique, fairly representative of the views of progressive ideologues of education, is that there is an inbred incompatibility between the structuring of the school curriculum on the basis of subjects, and child-centred education. At the secondary level this boils down to the contention that an integrated curriculum is more conducive to the promotion of pupil development in that it is said to be based on skills rather than knowledge acquisition. The accusation that the subject specialisms concerned have made no historical attempt to justify the existence of their subjects betrays, to say the least, an airy disregard of extensive literatures.

Another feature of the argument is that conspiracies are at work in boosting the vested interests of particular subject bodies (Goodson, 1983). It is not, however, made clear why it is contended that some bodies such as the Geographical Association and Historical Association have engaged in sinister political lobbying; while others (for example, associations of economists and other social scientists, progressive educationists, teacher unions, as well as organisations representing a whole range of environmental and other interests) are presumably conceived to have eschewed it. It may be that some have had a better case to make than others and made it more effectively!

Combating the arguments of progressives and conspiracy theorists is an alternative conspiracy theory. Thus Musgrove argued that subjects were not only intellectual but also social artefacts, and that a diversity of departmental interests in a school was important in averting an undue concentration of power: 'Only one man wins when you integrate subjects and dissolve departments – the man at the top.... Centralisation commonly entails standardisation and servility. Vitality lies in a vigorous, even defiant, pluralism' (1973, pp.8–9).

The phenomena of the 'senior management team' and the bureaucracies of 'faculty structures' have, since the time Musgrove wrote, come to haunt many schools, as those who once functioned as heads of subject departments spend less and less time in the classroom, now the job of an under-race of teachers. Arguably, if the more experienced and capable of geography teachers are moved into administration, the children will suffer. Of course, this may be as equally distorted a portrait of reality as the Kelly and Goodson stereotypes. Offering counter-stereotypes is, however, a start in provoking us to dig deeper into the issues.

Having made these points, those in geographical education and other subject areas must accept that there is a case for subject specialists to answer. Subjects such as geography cannot regard their place in the curriculum as sacrosanct, and must see their inclusion as contested.

At the primary level, the preferred progressive alternative to subject-based teaching and learning has been topic work, associated with the class-teacher model. This has been quite uncritically accepted as the essence of good primary practice, as is evident in the introduction to Tann's (1988) edited book on the subject:

'Good topic work is the epitome of all that is best in British primary education. It is an all-embracing way of working and one that is infinitely flexible.... The subject-based curriculum framework and objective testing would seem to be at odds with the topic way of working.' (p.1)

While Tann appropriately offers qualifications to the extent of insisting that the topic work must be good and teachers highly skilled and sensitive, there are a whole series of unexamined assumptions that topic work and subject-based work are incompatible, and that subject-based work is necessarily associated with a narrow didacticism.

Alexander (1984) is not alone in pointing out the paradoxes, contradictions and hidden agendas of progressive practice. Looking back into the elementary tradition of the nineteenth century, he observes:

'The class-teacher system...was the cheapest and most straightforward means of educating children to the minimal levels required...(In) the twentieth century...there was need to develop a conceptual framework for the practice of class teaching which...would support and sustain class teachers. Child-centredness...provided the best available ideology to meet the primary class-teacher's situation....' (p.14)

The basic paradox is, however, that the progressive rhetoric masks a wide range of practices, both within schools and as between schools. Thus:

'...mathematics is taught mainly as a separate area of learning...untainted by enforced conjunction with other curriculum areas...in acquiring "basic" status, areas of learning divest themselves of integrationist attributes...and become, to all intents and purposes, separate subjects....'(p.67)

On the other hand, it is argued that geography and history as distinctive modes of enquiry have no validity for primary and middle-school children, and also for the lower forms of secondary schools, and that what is necessary is an undifferentiated topic-based 'exploration' or 'discovery' of the environment (p.119).

Similarly 'art', argues Alexander, is subject-centred but 'aesthetic development' is child-centred:

'Which is more serious in its consequences for children's education, a view of "art" as a distinctive area of the curriculum involving induction into capacities to use and respond to the elements of colour, tone, form, texture, line and so on...or a view of "aesthetic development" as comprising no more than basic manipulative skills and a sense of pleasure? In terms of under-expectation and under-achievement, it could be argued, the child stands to lose more from the imposition of a narrowly conceived developmentalism than from a subject-based curriculum.' (p.27)

An important sub-text is that while archetypal progressives may argue that subject teaching threatens child-centredness, the tougher problem is for generalist class teachers. To quote Alexander again:

'...the erection of subject barriers poses the threat, because in this case the teacher may lack the requisite specialist knowledge...the only defensible curriculum is an integrated one with low "boundaries" or, where subject insecurity is greatest, an undifferentiated one which does not even admit the validity of subjects.' (in Pollard and Bourne, 1994, p.210)

Thus in areas where, subject-wise, primary teachers are relatively expert, as in the traditional elementary basics of English and mathematics, it seems a subject approach can confidently be practised. But this is not the case, for example, in the humanities and aesthetic subjects, seen as requiring a topic-based approach. Accepting the premise that subject divisions are not high priorities in the learning of younger children, can it be argued that some subject expertise in a range of subjects is unnecessary in the primary school? The serious under-expectation of the intellectual capacities of children as they develop towards the secondary phase, has been especially evident in the subjects beyond the basics, as HMI reports of the 1970s and 1980s have revealed (DES, 1978a,b, 1989a,b).

The debate between subject-centredness and child-centredness has therefore regularly been presented as a primitive polarisation. Similar arguments have been applied in the secondary sphere. Subject practice is caricatured in the integrationist literature as teacher-directed, the whole class similarly engaged in acquiring pre-digested information, geared to terminal assessment, with the personal and social development of pupils ignored. By contrast, the classic child-centred approach is presented as fully integrated throughout the day, with the individual child or groups of children evolving its or their own project study, following intrinsic interests, perusing primary resources, engaging in first-hand experience and experiment, and keeping a personal log of achievement for formative evaluation.

The stereotypes emerge from the old dichotomies of English education. On the one hand there were the extremes of the grammar school tradition, geared to the needs of able children wishing to pursue the academic route to higher education. This preoccupation resulted in the procedures of the sixth form being imposed on 11–12-year-old pupils. At the other end, a progressive kindergarten type ideology had taken hold, with 10–11-year-olds taught on the basis of methods originally evolved for infant children. The change from the primary to the secondary phase in extreme cases represented an unnerving discontinuity of pedagogic practice. Children were sacrificed to the tenets of competing ideologies.

What is particularly perturbing, a case of not letting the evidence get in the way of a dogma, is that many proponents of integrated topic work cite the pioneers of progressivism as justification for their ideological location. The historical evidence reveals, however, that it was almost universally not the case that the progressive pioneers objected to subject teaching. What they were against was the nature of the work being practised under the subject headings.

Thus, Matthew Arnold attacked 'capes and bays' teaching, but not geography in the curriculum as such (1869). Similarly the arch-English progressive and former HMI, Edmond Holmes, wrote about geography (and history) in a differentiated manner:

'Information as to the names and positions of capes and bays, as to areas and populations, and other geographical facts, is easily converted into knowledge of those facts, but it is not easily converted into knowledge of geography...in the absence of...the geographical sense, the possession...of geographical information cannot be converted into knowledge of...geography.' (1911, pp.90–134)

Above all, John Dewey was markedly unsympathetic to the kind of polarisation practised by many of his later disciples:

'How, then, stands the case of Child *vs* Curriculum?.... The radical fallacy in the original pleadings with which we set out is the supposition that we have no choice save either to leave the child to his own unguided spontaneity or to inspire direction on him from without...the value of the formulated wealth of knowledge that makes up the course of study is that it may enable the educator to *determine the environment of the child,* and thus by indirection to direct.' (1902, p.3)

While Dewey objected to subject matter being presented from above to children as a summary of adult experience, he regarded it at the same time as of prime importance in education, which he saw in a balanced way as having an intellectual and social as well as a child-centred purpose. Geography, like other subjects, must of course be properly presented:

'...the difference between penetration of this information into living experience and its mere piling up in isolated heaps depends on whether these studies are faithful to the interdependence of man and nature which affords these studies their justification. The function of historical and geographical subject matter...is to enrich and liberate the more direct and personal contacts of life by furnishing their context, their background and outlook....The classic definition of geography as an account of the earth as the home of man expresses the educational reality.... When the ties are broken, geography presents itself as that hodge-podge of unrelated fragments too often found...a veritable rag-bag of intellectual odds and ends.... Geography and history are the two great school resources for bringing about the enlargement of the significance of direct experience. Unless they are taught for external reasons or as mere modes of skill, their chief educational value is that they provide the most direct and interesting roads out into the larger world of meaning.' (1916, pp.210–3).

The Deweyan notion of subjects as resources for activity-based learning, rather than as artefacts subversive to child development, was evident in the work of the Liverpool Schools Council Project *History, Geography and Social Science 8–13*. The project distinguished usefully between 'subjects' and 'disciplines' and refused to be drawn into the subject versus child-centred polarisation. It regarded 'subjects' as timetable labels, which might or might not be used, and 'disciplines' as valued and indeed essential resources, to be deployed in inter-relationship in curriculum planning for the middle-school age range.

Forms of Curriculum Integration

It would seem important to build on the balanced thinking of the likes of Dewey and the Liverpool Schools Council Project, and provide a more differentiated analysis of the many different things that subject integration in practice can mean. Warwick offered such an analysis as curriculum integration was spreading into secondary schools in the period of comprehensivisation. Part of the justification for such integration was that less able pupils in the secondary phase were, like primary pupils, seen to be more appropriately catered for by an integrated approach: social studies rather than geography and history (Graves, 1968).

Warwick's Classification

Warwick (1973) identified various forms of integration, classified according to the degree of structuring that was present:

1. At the unstructured end of the spectrum, the approach was one 'in which the subjects as such do not make an appearance at all and in which children are free to develop their own individual and group approaches to large **open-ended themes'** (p.2), or topics.

2. The first step towards a more formalised position he termed **theme teaching**. Here the subjects were present in the background to give shape to the integrated scheme. The situation was a lightly structured one with balanced subject inputs. In a later terminology, this could be defined as a **focused topic** approach.

3. **Faculty teaching** described the organisation of individual subjects into faculties such as 'humanities', 'creative arts', etc. The contribution of each subject was, theoretically, balanced, and within each faculty a 'common-core' syllabus evolved through discussion. One of the subject departments concerned might take a leading role. The approach was more formal than the first two categories, with content and direction controlled by the teachers. Team teaching procedures could be used. The timetable was blocked, formally recognising the existence of an integrated arrangement.

4. Related or **combined studies** was used to refer to situations in which there was no formal timetable recognition of faculty groupings. History, geography and other subjects continued as timetable labels. 'Integration' took place to the extent that subject heads found out what others were doing, when they were doing it, and the order in which they were doing it. Cross-referencing was to be sustained and joint lessons or exchanges of teachers if possible arranged.

Warwick regarded combined studies as 'a legitimate first step to any closer liaison' (p.3). It reflected a pragmatic solution, involving a limited degree of co-operation between staff in organisational terms, and a complementary rather than a truly integrated approach to themes in the structural sense. A major disadvantage of combined studies approaches was that the viability of the integration was left to chance or, at best, to informal staff-room discussions. Even in more formalised schemes, what might seem to be integrated on paper might well not be in the experience of the pupil, left to do her or his own integrating. An example of a combined study approach to a topic on 'The Sea and Norway' is illustrated in Figure 4.1.

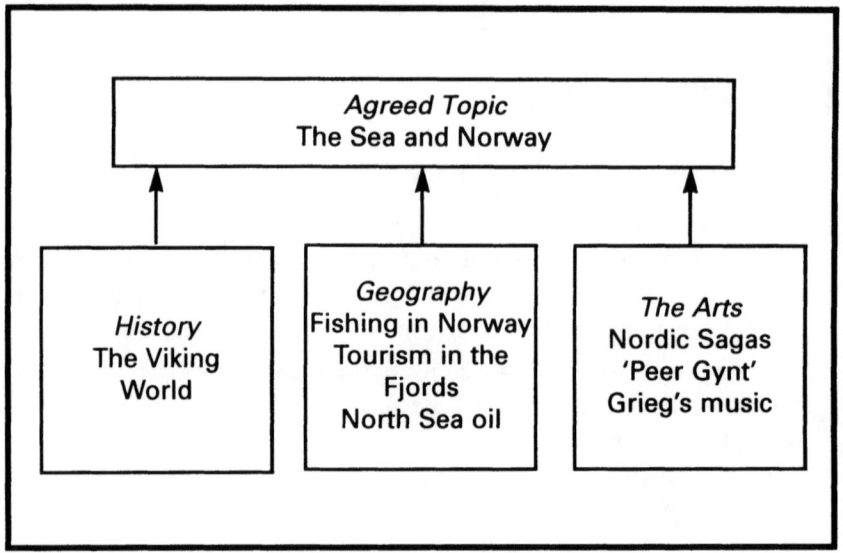

Figure 4.1 Sub-topics covered by separate disciplines

Inter-disciplinary Enquiry

Within the structure of knowledge broad fields of experience can be recognised. Interpretations of what these fields may be differ, but in practice they would normally include the sciences, the humanities and the creative arts. These have often formed the basis for faculty structuring. But Pring (1971) raised the question of what integrating basis there is in, for example, the humanities?

> '...we are bound to ask in what way does this theme or this area of study itself give a basis for structuring or organising knowledge other than in the worked out disciplines of thought. How does a child classify what he observes if not within the conceptual scheme that owes its formal structure to the basic and differentiated forms of knowledge.' (pp.133–4)

There are also complex connections between the different areas of knowledge. In addition to concepts intrinsic to particular disciplines, there is a range of inter-disciplinary connections, especially in areas such as the social sciences, which share many concepts. True inter-disciplinary enquiry is concerned with interaction rather than mere juxtaposition. At the organisational level it implies detailed co-operation between staff in which agreed statements of objectives are spelled out and course structures jointly planned. The purpose of the exercise is to bring together the concepts, principles and methodologies of different disciplines in an attempt to come to grips with problem issues which could not be grasped as effectively by individual disciplines alone. A preliminary task is to identify 'problems' that require such study. These need not be solely real-life social and environmental problems, though these will figure prominently in such schemes. In the real world of planning, for example, a team of researchers will be brought together to exercise their skills on, say, urban traffic or rural conservation issues. In the classroom, simulated problems of this kind can usefully be introduced. Links have to be established at two levels, as illustrated in Figure 4.2 through the unifying theme 'Using the North Sea's Resources Wisely', which is a focused topic here given a value-laden, problem orientation.

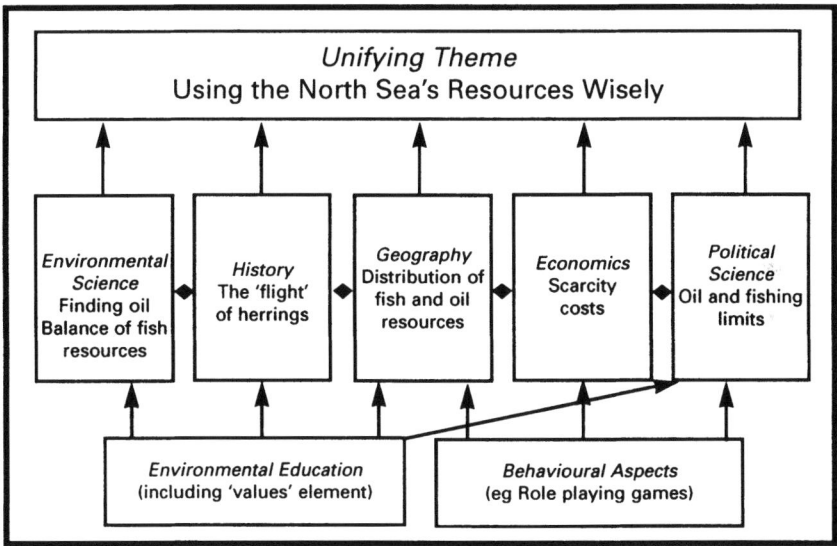

Figure 4.2 Disciplines organisationally linked

An inter-disciplinary approach is self-evidently needed in any issues-based thematic framework. It is difficult to think of any real-world issues that can be encompassed by the approaches of one particular subject, still less by a loose, discipline-free topic approach. This is a point to be discussed further in Chapter 9. Major problems relate to the individual expertise of teachers and the complexities of drawing the necessary range of skills together in the particular school context, to produce a coherent and intellectually honest set of solutions.

Geography as a Curricular Pivot

Geography has long been presented as having a pivotal role in the curriculum, bridging the humanities and the sciences. Among the early pioneers of geographical education as we know it today, Geikie, as an earth scientist, was especially interested in the links between geography and science (1879). Within its broad aims of studying the interaction of peoples and their environments over the earth's surface:

'geography, wrote Geikie, comes frankly for assistance to many different branches of science.... It brings to the consideration of their problems a central human interest, in which these sciences are sometimes apt to be deficient; for it demands first of all to know how the problems to be solved bear upon the position and history of man and of this marvellously ordered world.... Geography freely borrows from meteorology, physics, chemistry, geology, zoology and botany; but the debt is not all on one side.' (p.423)

At the time Thomas Henry Huxley was promoting a subject which linked geography and science in schools, namely physiography. Physiography was a key subject, in Huxley's view, in that it demanded attention to the natural world, first-hand experiences, and scientific processes (1877, pp.vi–vii), and helped to steer the curriculum away from the abstractions of science and geography, as they were then taught. For a time, the subject achieved some success, and formed the basis of early field work, as reflected in Cowham's text (1900).

From the side of geography, there were others who were to point out the potential linkages. B.F.D. Harris (1934) argued that geography's most important role in the curriculum was correlative, helping both to unify the curriculum and to present points of contact between school subject studies, and between these and actual life. On the scientific side, geography teachers should help their pupils to get 'inside the skin' of the scientist. As we have seen, the quantitative revolution of the 1960s was to reinforce the importance of such connections (*see* Chapter 2).

Throughout his academic career, Mackinder drew attention to the pivotal position of geography (1921) as between the humanities and the sciences. This he saw as a reflection of the essence of geography, which demanded the connection of its political (human) and physical tissue:

'It postulates both scientific and humane knowledge. No one can appreciate geographical correlations without some mathematical, some physical, some economic, and some historical knowledge. Geography is essentially a mode of thought which has its scientific, artistic and philosophical aspects. If our aim is to give unity to the outlook of our pupils, and to stop that pigeon-holing of subjects in their minds which has prevailed in the past, then geography is admirably fitted as a correlating medium. It may very easily be made the pivot on which other subjects may hang, and hang together.' (p.382)

In a memorandum to the Board of Education in 1934, the Geographical Association similarly denied the subject's image as one 'concerned mainly with memorising names or furnishing footnotes to history', but as taking a synoptic view on world circumstances by exploring connections and mutual exchanges between its regions. In doing this, the

'fields of observation included both the sciences of observation of nature and the humanities, and geography thus claims a special educational importance in that it links these two disciplines and, in addition, gives abundant exercise to the powers of descriptive writing, artistic delineation, and individual enquiry.' (pp.47–51)

Translated into school terms, the movement towards curriculum integration in the 1970s resulted in a large number of variants. The presence of geography in most of them revealed its nodal position in the curriculum:

- *Environmental science* – The physical elements of geography, with ecology (or biology), geology, meteorology, pedology (soil study) and so on.
- *Environmental studies* – Both physical and human aspects of geography, local history, ecology (or biology), and perhaps a social science component (for example, archaeology).
- *Social studies* – Largely human aspects of geography, with history and the social sciences. More specific sub-categories were offered in the 1980s, including **peace studies**.
- *Humanities* – The human aspects of geography, with history, religious knowledge and English.
- *Outdoor pursuits* – The physical aspects of geography, with earth sciences, ecology (or biology) and physical education: a later link, through orienteering, is with physical education and mathematics.
- *Area studies* – Not so much local studies (*see* environmental studies) as studies focused on large-scale units such as **European** or **American studies**, drawing on regional geography, history, literature and other creative arts, languages and the social sciences. During the 1980s further variants emerged, such as **world** (or **global**) **studies**.

Unlike the position of integration in the primary phase, that in the secondary in the 1970s was heterogeneous. Some schools ran a mixed economy, using integrated studies in the lower forms, with students proceeding to undertake GCEs and CSEs in particular subjects. Others plumped for an integrated, yet others for a subject-based curriculum. In a period of autonomy, individual schools and even individual teachers could make the choice. There was no sense of a general entitlement, either one way or the other.

Achieving Coherence in an Inter-disciplinary Framework

In achieving coherence, three problems that are posed are of balance, sequence/progression and focus.

The Problem of Balance

The broad balance between elements in the structure of knowledge is met with to some degree in traditional curricula. The current rhetoric refers to the importance of breadth and balance in the curriculum. In practice, balance may be more difficult to achieve in integrated curricula where there is a more divergent spread of content, as well as an even more divergent spread of views, to be incorporated. In the actual process of curriculum planning, negotiation proceeds on a personal or group basis. Where there is a subject-based timetable, with agreed time allotments, at least a quantitative balance is achieved. In less tight integrated-curriculum planning, decisions can be affected by the personal powers and commitment of one individual, by a clash of personalities and/or by an unbalanced range of subject expertise across the group. Many integrated schemes of work have been unbalanced by a dominant historian over-emphasising the temporal, or a geographer the spatial dimension. Similarly teachers lacking in confidence in particular areas may be quite happy seeing these areas marginalised. This has been a particularly serious problem in the primary school, where idiosyncratic choices have often reflected the particular interests of headteachers. For such reasons, a balance across the curriculum may be prejudiced.

The Problems of Sequence and Progression

The problem of sequencing material is already a difficult one in geography, where the core concepts and principles often do not build on each other, are characteristically of parallel importance, and have strongly overlapping tendencies. The task becomes extremely complex when subjects such as geography are joined with similar subjects. Thus Graves (1968) noted that:

'developing a worthwhile course on IDE [inter-disciplinary enquiry] lines must be an extraordinarily difficult task, since the topics chosen must not only enable the dove-tailing of various disciplines in some coherent pattern, but each succeeding topic chosen must build upon the previous topic in such a way that principles and concepts learnt are gradually enriched and developed.' (pp.392–3)

There are also discontinuities between the methodological skills which each separate subject regards as distinctive. It is more straightforward, for example, to build a logical sequence of map-reading skills into a geography curriculum than into an integrated one. One of the most widely advocated methods of sequencing is Bruner's **spiral curriculum** idea in which major concepts are revisited at different stages, each later stage involving a greater degree of complexity (1960). This is, of course, difficult to achieve even in a subject-centred curriculum. There would seem to be a need for an interlocking set of spirals in a broad social studies arrangement. The problem is self-evidently associated with the issue of progression, which will be returned to in Chapter 6.

The Problem of Focus

As we have noted, the choice of a broad topic such as 'The Sea' or

'Homes' does little in itself to prevent an arbitrary selection of content. Even where certain aspects are emphasised through the introduction of disciplinary criteria, there may still be a lack of focus. Hence a project on 'Homes' can include 'Roman settlements' from history, 'igloos and kraals' from geography, baroque architecture in art and craft, 'safety in the home' in home economics, and 'homes in Victorian novels' in English literature. Other criteria need to be brought to bear to enable a coherent choice to be made.

To illustrate the problem, let us look at two topics in Tann's book: one based on 'France' and the other on 'Journeys' (Figures 4.3 and 4.4). In the first, the topic is ostensibly geographical, but the approach is designed to peripheralise subject input. While a whole range of subjects is picked out on the diagram, there is no implication that these subjects have distinctive conceptual frameworks to offer, as essential resources for developing understandings. It is useful to try to analyse this structure on the basis of various criteria. What leads us to believe in Figure 4.3, for example, that more attention is being paid to connection, rather than to collection? Taken at face value, this brain-storming offers a *carte-blanche* to attempt almost anything, and a potential coverage that makes even the pre-Dearing National Curriculum seem under-weight. In Figure 4.4, an expert

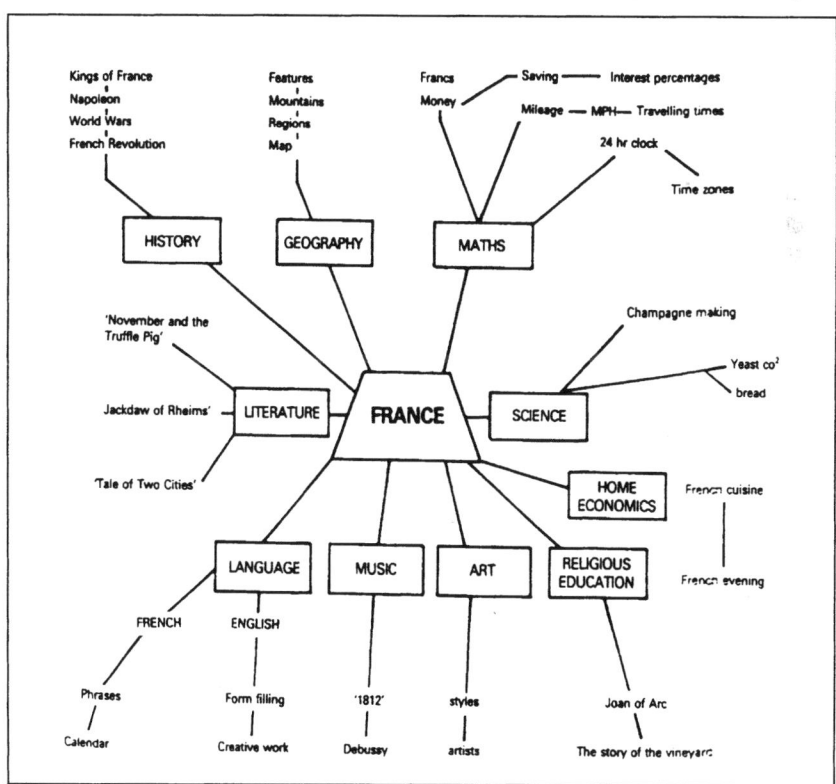

Figure 4.3 Topic web for work on France

54

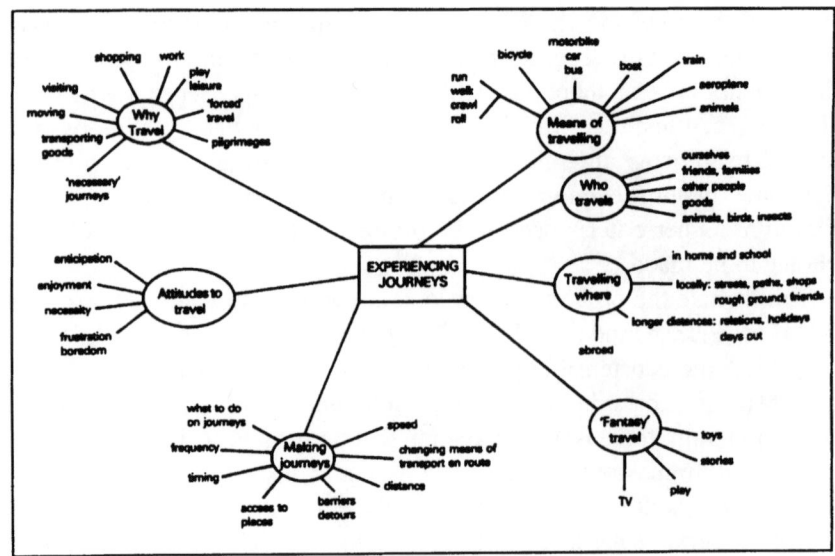

Figure 4.4 Central elements for study in a journey's topic

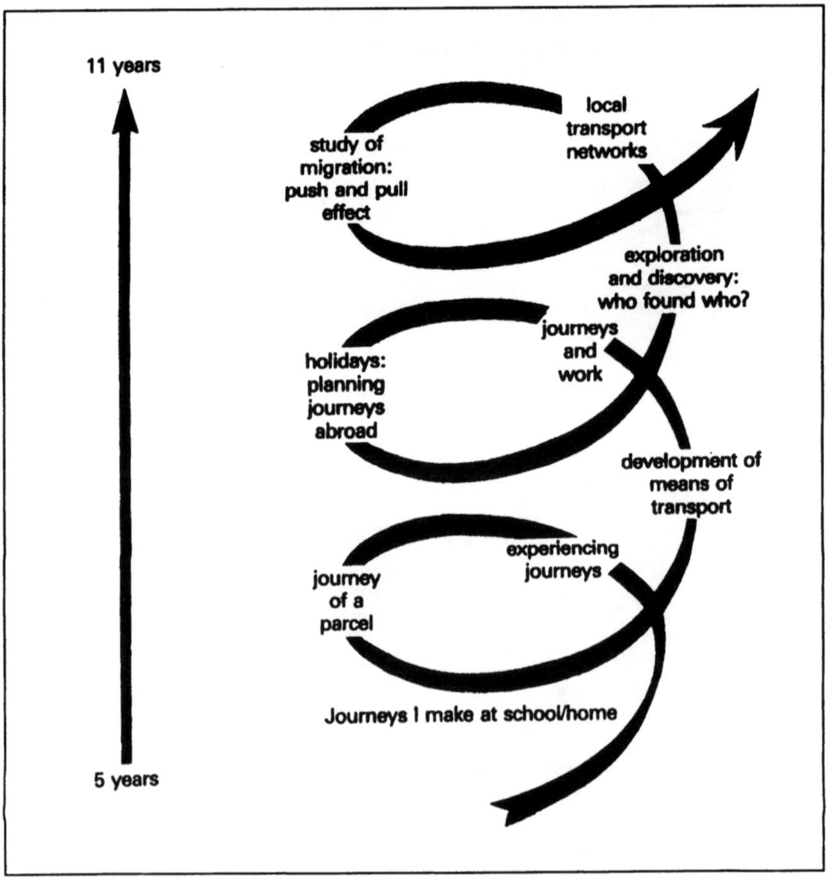

11 years

study of
migration:
push and pull
effect

local
transport
networks

exploration
and discovery:
who found who?

journeys
and
work

holidays:
planning
journeys
abroad

development of
means of
transport

journey
of a
parcel

experiencing
journeys

Journeys I make at school/home

5 years

Figure 4.5 Journey topics in a spiral curriculum

geographical educationist, Catling, who at the same time supports progressive approaches, seeks to bring coherence by working on key questions and ideas associated closely with the topic and drawing on a more limited subject range. This is indeed is a useful example of a focused topic, which gains coherence from the fact that connections are being sought. The author illustrates also how his focused topic can be revisited in a spiral curriculum framework (Figure 4.5).

Two useful ways of achieving focus are through the working out of area-based schemes, more distinctive to geography; or concept-based schemes, based on principles or key ideas, which can be geographical but may be interdisciplinary. Into the latter, the resources of particular subjects are permeated.

Area-based Schemes

Area studies of different types can help to achieve focus, unless the scale of the study is too large, like 'France' above. In these circumstances, the areal component can be as diffuse as any broad topic. Local studies at various scales, even detailed site studies, are a good way of focusing the contributions of particular disciplines. Many integrated schemes at all stages of schooling are based on the stimulus of local studies in the field. Yet a number of well-known educationists have been disturbed by the potential of such schemes for engendering parochialism. Taba (1962) saw local studies as 'ethnocentric' when the prior need is to develop a 'cosmopolitan' viewpoint (p.273). Bruner (1966) was similarly sceptical of studies based on 'the familiar world of home, the street and the neighbourhood' and 'the friendly postman,' and draws attention to the difficulty of generalising from the familiar (p.93).

As we saw in the previous chapter, detailed area-based initiatives are present in sample/case/locality studies. Here a **concentric approach** is a useful foil, the principle being to use the locality as a base from which to move out over the world concentrically, say from local to regional, then to national, then to global levels, widening the focus on a particular topic or concept. Masterton (1969) illustrated this principle effectively in a broader environmental studies context showing, for example, how ideas about mountain environments and their use can be focused and sequenced by starting with the vicinity of the local hill or mountain, thence moving outwards to more distant places and broadening scales. The idea of a concentric approach must be correctly interpreted. What it should not mean is that local studies are covered in the primary school, and national and international in the secondary. It means rather moving out reasonably quickly from the local area, theme by theme, at any particular stage, including the infant.

Concept-based Schemes

An alternative to the topic- or area-based study is the more fully organised method of employing broad concepts as a basis on which work can be structured in a truly inter-disciplinary way. Bernstein (1971) regarded as necessary for real integration:

> 'the presence of a relational idea, a supra-content concept, which focuses upon general principles at a high level of abstraction.... Whatever the relational concepts are, they will act selectively upon the knowledge within each subject which is to be transmitted.' (p.60)

Such concepts are likely to be expressed at high levels of generality. They may be in the nature of the broad key concepts such as similarity and difference, continuity and change, as we have seen, or principles (key ideas), or key questions, as used by the various Schools Council projects discussed in Chapter 3. Several examples may be used to illustrate differing interpretations of this type of process. First, Bruner's well-known and anthropologically orientated *Man: A Course of Study* is organised round three recurrent questions. What is human about human beings? How did they get that way? How can they be made more so? In pursuit of these questions, five great 'humanising forces' function as over-arching themes, namely tool-making, language, social organisation, the management of man's prolonged childhood, man's urge to explain the world (1966, pp.74–5). The interpenetration of these themes is emphasised. The contributions of the social and behavioural sciences, the humanities and the natural sciences, demonstrate that the ambition of the scheme is to furnish a near-total curriculum.

It is suggested that the use of statements of principles and generalisations is a helpful way of structuring inter-disciplinary content, as well as providing a source of objectives. The following example is by way of a prelude to fuller discussion of these matters in Chapter 5. An attempt to resolve the problem in a setting wider than a particular subject area is explored in connecting the theme of 'barriers, divides and frontiers' with an area studies framework, here European studies.

Key Ideas in European Studies: 'barriers, divides and frontiers'

1. Conflicts between groups and nations are associated with division: for example, social divides, economic barriers and political frontiers, among other things.
2. Political frontiers reflect divisions between nation states and isolate and protect people of similar national identity.
3. Such frontiers may reflect natural geographical obstacles such as rivers, marshes and mountain ranges.

4. Artificial barriers may be built as a substitute or reinforcement for natural barriers: e.g. Hadrian's Wall, the Maginot and Siegfried Lines, the Berlin Wall.

5. Religious differences have an ideological basis, and are the product of long-standing historical contexts, though with less well developed geographical expressions than national frontiers, e.g. north and south Germany, Ulster and the Republic of Ireland.

6. Ethnic divides result from migration of peoples and are often a cause of serious internal conflicts within otherwise ethnically homogeneous states, e.g. Jews in Europe, displaced persons, migrant workers.

7. Linguistic divides are serious barriers to communication, both between and within states: but compare Belgium and Switzerland.

8. Social divides are common within all countries, and operate at individual and group levels (social class divides): they also have ideological overtones (e.g. capitalist versus socialist systems).

9. Economic barriers reflect different resource bases, and the need to protect home industries and markets from overseas competition. Countries may join to erect communal barriers against outsiders, e.g. the European Community. Economic differences also operate at group and individual levels.

10. Cultural differences: in many ways Europe has a 'common culture', but many regional and national differences occur in terms not only of language, but also of architecture, customs and life-styles.

11. National, cultural and social differences lead to the formation of 'inferior images' or stereotypes. Other significant concepts in this area include patriotism and propaganda.

This provides only a general inter-disciplinary framework for a large cross-curricular unit. Depending on the scope of the unit, each of these major principles needs to be broken down further. Then each of the sub-principles requires further unpacking before lesson units emerge. The way the disciplines of knowledge should be permeated needs to be carefully delineated. It is clear that a mono-disciplinary approach to, say, frontiers, would be more straightforward. But would it have as much credibility as an issues-based theme?

The Role of the Teacher

The imposition of integrated schemes can pose serious problems for teachers who lack confidence, and genuine understanding of the new arrangement. The commitment of teachers to integrated codes is a crucial factor in whether or not they are successful. As earlier implied, the integrated scheme may well be imposed from above and, as a top-down exercise, not have the benefit of what the current terminology calls the 'ownership' of the teacher. Bernstein (1971) spelt out the issues and the tensions:

> 'There must be consensus about the integrating idea and it must be very explicit.... Integration codes...weaken specific identities...[and] require a high level of ideological consensus and this may affect the recruitment of staff. Integrated codes at the surface level create weak or blurred boundaries, but at bottom they may rest on closed explicit ideologies. Where such ideologies are not shared, the consequences will become visible and threaten the whole at every point.' (p.64)

Teachers used to working under traditional schemes require a process of socialisation into the integrated code, for the procedures of the new code will need to be as much internalised as those of the old. It is well known that integrated schemes in both intellectual and organisational terms are more difficult to work than traditional schemes. As Bernstein writes:

> 'The collection (i.e. subject-based) code is capable of working when staffed by mediocre teachers, whereas integrated codes call for much greater powers of synthesis and analogy, and for more ability to both tolerate and enjoy ambiguity at the level of knowledge and social relationships.' (p. 65)

The National Curriculum, with its subject-based structure, may be thought to have administered the *coup de grâce* to integrated schemes of work. This is not the case. Whatever its disadvantages, the National Curriculum has made a much larger number of teachers aware of what is going on in other subject areas, and of the potential links between them. The term 'integration' has indeed largely been made redundant in the current reference to cross-curricular activity. Much thought has been put into cross-curricular guidance. If there is evidence that elements at the Department for Education regard cross-curricular activity as peripheral, many teachers do not. For a start, they appreciate that in an over-crowded curriculum the cross-curricular links can be audited for subject content with potentially great time- and energy-cost benefits. The subject of cross-curricular issues and the National Curriculum will be returned to in Chapters 9 and 10, where the crude polarities of integrated versus subject based frameworks will be avoided, and discussion will centre round the more subtle differences between subject-permeated issues based schemes, and issues-permeated geography-based frameworks.

GEOGRAPHY AND EDUCATION

CHAPTER 5

Curriculum Theory and Geography

Introduction

The contention that elements of good practice have always been present has already been made. What has yet to be achieved, it has been suggested, is the realisation of a balance of good practice, as between geographical, educational and social dimensions. In secondary education, for example, whether under the dominance of the regional geography paradigm of the old grammar school system, or in the issues-based approaches of the comprehensive, too little thought has been given to other aspects of good practice, not least those supported by curriculum theory.

Various arguments have been used to demote curriculum theory as an important underpinning of actual curriculum planning. One is the down-to-earth assertion that such theory is airy-fairy and divorced from the realities of the classroom existence. Another frequently expressed opinion is that of the progressive lobby, characterising curriculum theory as inevitably bound up with narrow, prescriptive and constraining objectives statements, predetermining outcomes and leaving nothing to pupil initiative. This latter view is particularly important to combat in the light of the fact that, if any curriculum theory is present (if so it is implicit and not explicit) in the National Curriculum, it is of a narrow variety.

Curriculum planning by objectives has also been termed **rational curriculum planning**. It is perhaps a sign of the times that logically so beneficial a term, in any educative context, as 'rational', can be inverted into a label of abuse. We need to acquire a more differentiated understanding of what curriculum planning by objectives means before we dismiss it too easily.

Varieties of Curriculum Planning by Objectives

Educational writers of the 1970s were responsible for many false chartings of the history of curriculum planning. Most correctly attribute systematic curriculum making to American educationists, but variously suggest the approach was pioneered by Tyler and Taba (1930s–1960s) or Bobbitt (1910s and 1920s). In fact, Tyler and Taba can be seen as the third generation of American experts in this field, and Bobbitt as belonging to the second. The movement goes back further, however, at least to the curriculum planning initiatives of William Torrey Harris, superintendent of the St. Louis school system in the 1870s. By the 1890s prestigious school superintendents and professors of education had formed active educational pressure groups, implementing their ideas through influential curriculum-making committees (*see* Marsden, 1991). Lamentably, the whole rich tapestry of nearly 100 years of curriculum development has been distorted by attention being directed to a small number of scapegoat figures, notably Bobbitt and Tyler.

It is perhaps an unfortunate aspect of American education that many state authorities subsequently seized upon the more mechanistic aspects of post-war curriculum planning by objectives in implementing change. On the basis of such examples, some suspicion of the American experience may be justified. They have given curriculum planning by objectives a bad image.

What are curriculum objectives? They are by definition more limited in their scope than aims, requiring more precise statement. They serve in part to translate the broad guidelines provided by statements of aims into what it is possible to achieve. The purposes for which the objectives are formulated will help to determine how specific they need to be. They can, for example, be related to three levels of educational provision:

- overall curriculum planning for the school;
- overall syllabus planning within an area of study; and
- planning at the level of short course or lesson units.

The first covers the strategic planning, and long-term objectives should be specified for the wider, curricular and extra-curricular provision by the school, within the parameters laid down in its overall aims. Schools are of course now required to offer more than rhetorical aims statements, and provide a basis for justifying what they are doing by spelling out general objectives. Within each area of study, departments or faculties are expected to have overall syllabus planning objectives. Those related to course or lesson units will be more specific, and it is these that are more likely to be narrowly defined. Objectives might equally be viewed in terms of what the learner is expected to have achieved by the time s(he) leaves school, by the end of a particular stage of schooling, or by the end of a course unit. But as we shall see, it is important to specify **process** as well as **outcome** objectives.

Operational or Behavioural Objectives

The most precise and primitive statements of objectives are those termed 'operational' or 'behavioural'. They refer to pre-specified observable changes in pupil behaviour. The word 'behaviour' is not, of course, being used in this case in the sense of 'conduct', but refers rather to cognitive behaviour. The Bobbitt approach (1918) was based on the sort of statements of objectives more appropriate to the training of mechanics or cooks, where detailed behaviours could be specified in sequence and be measured by ticking boxes. It is exemplified in such targets as 'State the behaviour in a way that somebody else could count it' (Neisworth *et al.*, 1969, p.5). Perhaps the clearest exposition of an unbridled post-war behavioural objectives approach was provided by Mager (1962), who insisted that statements of objectives should conform to the following criteria:

1. They should be stated in performance terms that describe what the learner will be DOING when demonstrating his achievement of the objective.
2. The language in which the statement is expressed should communicate the same thing to different people, and not be open to a wide variety of interpretations. In any statement of objectives, infinitives such as 'to identify' are more satisfactory than those such as 'to understand'.
3. The conditions under which the behaviour is to occur should be stated. The materials to be used, and even the time allowed for the task, should be specified.
4. The lowest level of acceptable performance should be stated.
5. A separate statement should be associated with each objective, as the more statements made, the more likely they are to be clear in their intent.' (pp.10–12)

Such pre-emption of any creative educational experience clearly justified the complaint of Charity James (1968) at the time:

'This is the language of assembly-line processing of products.... The underlying image is of the factory rather than the consulting room or the school...the whole exercise of thinking in terms of block objectives...is a system for disregarding individual needs, a system for creating conformity.' (p.88)

Following Mager-type prescriptions, it can be inferred that there will be a tendency for teaching to be didactic and instrumental. Referring back to the 'payment by results' experience, with all its procedures statutorily laid down, the teacher was always likely to 'stick to the rule book' and close up opportunities for creative and spontaneous work. Another possible and even likely outcome is that behavioural objectives will be too closely associated with easily measurable and relatively trivial content. It is easy to specify, for example, that at the end of a unit of work on South America the pupil will be able to list the main exports of each country. It is more difficult, however, to identify higher level outcomes relating to problem-solving activities on, for example, issues of social geography, with such lack of ambiguity. Thirdly, the highly specific objectives of educational activity in the syllabus as a whole, and within subjects, will be too numerous for them all to be stated conveniently, a point well illustrated in the multiplicity in the statements of attainment in the first version of the National Curriculum (*see* Chapter 11). The position taken here is that Charity James was right in her condemnation of behavioural objectives of the Mager type. Detailed operational objectives of this type should have little presence in good practice, and unfortunately they appear to be creeping back in so-called competency-based appraisals, based on narrow performance criteria. At the same time, it must be emphasised that objectives specification need not be of this narrow type.

Differentiated Views

A more differentiated view of objectives would cover process as well as product objectives. A facile view would be that the former are necessarily good and the latter bad. It must be recognised however that catechetical techniques and rote memorisation are processes, leading towards the product of regurgitation of factual information. As has been characteristic of the polarisation of educational discourse over the last 30 years, the behavioural objectives extreme, which does concentrate on products or outcomes, has been widely categorised as the norm rather than as a polarity.

To repeat, specification of narrow behavioural objectives is just one variant of rational curriculum planning. Supporters of the broader view would counter some of the objections by stressing, in the first place, that far from leading to trivialisation of content, the identification of objectives clarifies which particular objectives are trivial and which are not (Popham, 1968). In this way trivia are brought into the open and can be kept under scrutiny. At the most straightforward level, for example, it is possible to specify beforehand that only a small proportion of the marks for a particular unit of work will be allotted for the recall of facts.

Prespecification of outcomes does not necessarily imply that the means to achieve them should all be preconceived. There are many routes to the same destination. There is no need, for example, to stifle the spontaneous use of opportunities for open-ended discussion which may crop up during a lesson. The delineation of process objectives can make this explicit. In addition, the unanticipated outcomes of teaching and learning can be assessed as well as those prespecified. The latter might in certain instances tend to form the essential core of what all children would need for later progress. Over and above this it would not be vital to tie the work to the originally agreed objectives. The use of a wide range of assessment techniques (*see* Chapter 7) makes possible the appraisal of a whole spectrum of intellectual skills. To reinforce an earlier point: one of the necessary refinements of the assessment procedure is the use of a test specification to control any tendency to over-emphasise less important educational activities.

In meeting the criticism that the behavioural approach entailed an unacceptable proliferation of objectives, Tyler (1964), quite wrongly tarred with the same brush as Bobbitt, argued strongly that clarity was a more important criterion than extracting the last ounce of specificity. The question of the degree of generality at which statements of outcome should be presented was discussed by Tyler in far more sophisticated terms than those of the extreme behaviourists. Two important suggestions emerged:

- that concepts and principles of particular disciplines form a more manageable check-list for the specification of objectives than descriptive content;
- that the level of generality at which objectives are expressed is a crucial factor in determining whether they can provide a useful structuring principle, without the participant becoming bogged down in a multiplicity of highly specific content statements.

Skilbeck's Compromise

It must not be assumed that all educational writers of the 1970s and 1980s dismissed curriculum planning by objectives. The practical assistance which a middle-of-the-road objectives approach could provide was conveniently summarised by Skilbeck (1971):

1. A wide range of content and materials present themselves to the teacher. A clear specification of worthwhile objectives provides criteria of choice. Significant content can be given precedence over trivial.

2. Important variables relating to the pupil are made explicit, including an assessment of his state of readiness, an examination of the possibilities of sequencing material according to principles derived from learning theory and a recognition of the need to individualise learning to meet the needs of particular pupils.

3. Detailed feedback, of use to teachers and pupils, is made possible.

4. In so far as the teacher is concerned, the approach:

   (a) Presupposes the sort of critical thinking from the teacher that in turn is expected from the pupils.

   (b) Helps to make explicit the success or otherwise of the teaching. This information must, of course, be viewed with caution, bearing in mind the variable nature of the objectives adopted and the varying situations in which different teachers find themselves. Trivial objectives are relatively easy to achieve, and some children are easier to teach than others, for example.

Bruner's Structures

Though not offering a taxonomy of objectives, another strong advocate of using an enlightened and rigorous structuring of the curriculum was Jerome Bruner (1960). Apart from the pleasure it could give, he saw as a central goal of learning the promotion of transfer of training. Various skills, principles and attitudes were necessarily acquired in school but, to be worthwhile, must later be applicable and of service in life situations. The training must therefore be transferable. In essence, this 'consists of learning initially . . . a general idea, which can then be used as a basis for recognising subsequent problems as special cases of the idea originally mastered' (p.17).

Such continuity of learning thus required mastery of the structure or fundamental ideas behind the subject matter. These fundamental ideas related to the generalisations, principles and concepts, and their interconnections, of the disciplines of knowledge. Teaching for structure was seen as having advantages for motivation, in that a subject cannot be exciting if its principles are not grasped, and particularly for cognition, since grasp of the structure makes the detail more easily remembered. Without going into the historical details, these ideas reflected the more balanced thinking of many members of the inter-war American curriculum planning movement, and were to be reinforced, as we shall identify below, in the work of notable American psychologists of education. They provided the basis, as Tyler suggested, for statements of objectives at a sensible level of generality.

Bloom's Taxonomy of Cognitive Objectives

The idea of rational curriculum planning by objectives is most usually associated with the development of thinking skills, in technical language, the **cognitive domain**. Bloom's taxonomy (1956) was a hierarchical scheme based on six umbrella categories, the so-called higher-level ones subsuming those below. The presumption was that the higher level skills could not be acquired before the lower. The detailed categories were:

1. Knowledge
2. Comprehension
3. Application
4. Analysis

5. Synthesis

6. Evaluation.

Each of these begat sub-categories. The scheme was enormously influential. It was also over-prescriptive and suspect in some of its basic assumptions, such as the idea that the categories were in hierarchical association. It was also ambiguous in its terminology. One of the disturbing aspects was not just the ambiguity but also the apparent low status ascribed to knowledge. Thus knowledge in its narrow sense can be thought of as the recall of basic information. On the other hand, it would be facile to attach to someone the credit of having a true 'knowledge of geography' that did not have an understanding of its approaches and explanatory frameworks, as will be explored later. The Bloom scheme none the less contained the important core idea that the development of thinking skills should be differentiated, and that each category was important. It demanded cautious but respectful scepticism, rather than the uncritical enthusiasm with which it was sometimes disseminated, or the glib condemnation heaped upon it from a different ideological extreme.

Ausubel's and Gagne's Schemes

The powerful influence of the Bloom taxonomy tended to obscure the presence of other schemes of the same period. Some of these were also based on hierarchical principles. The Ausubel (Ausubel and Robinson, 1969) and Gagne (1965) schemes were both associated with developmental learning theories, which would seem an appropriate start for planning cognitive objectives.

Ausubel's scheme can, in fact, be matched quite closely with Bloom's cognitive domain:

1. *Rote learning* formed the lowest level category, which Ausubel equated with Bloom's 'knowledge' category. In favourable contrast to Bloom's taxonomy, where 'knowledge' as information loomed large, Ausubel did not regard rote learning, in itself, as a worthwhile educational objective: rather as an extremely inefficient method of learning.

2. The next level was *meaningful learning*, presupposing the acquisition of concepts and principles, and more or less matching the Bloom 'comprehension' category.

3. *Application* was a category common to both schemes, and was defined by Ausubel as applying a principle in new circumstances in a fairly direct way. He saw it as difficult to distinguish from the category below.

4. *Problem solving*, which Ausubel defined as a more complex process than 'application', and involved the student in identifying and transforming relevant principles to achieve a desired result, namely, the solving of an intellectual problem.

5. Ausubel's highest category was *creativity*, which demanded the use of relationships in the student's mind to achieve a unique end product. Here the principles, concepts and strategies brought to bear by the student had not previously been taught as specifically relevant to the task in hand.

Gagne's hierarchy was made up of eight 'types of learning':

1. signal learning;
2. stimulus response learning;
3. chaining;
4. verbal association;
5. multiple discrimination;
6. concept learning;
7. principle learning;
8. problem solving.

The first three covered pre-school stages of learning and are not considered here. **Verbal association** and **multiple discrimination** could roughly be matched with rote learning, though this admittedly is an over-simplification of Gagne's definitions. In **concept learning**, the learner shows the ability to make common responses to classes of stimuli (for example, flowers) which differ widely from each other in physical appearance. The acquisition of concepts presupposes the ability to discriminate, to allot objects to particular classes, in the most simple form, **to distinguish exemplars from non-exemplars,** bearing witness to the power of contrast. Verbal definition was not enough. Concept formation was thus seen as a vital early stage in the discovery of meaning.

Principle learning, a stage on from this, involved the linking of two or more concepts, a process referred to by Ausubel as 'proposition learning'. Gagne's final category, **problem-solving**, demanded the ability to combine concepts and principles and apply them in new situations. These higher level activities were valued both by Ausubel and Gagne as cognitively efficient. 'A "higher-order" principle resulting from an act of thinking appears to be remarkably resistant to forgetting' (Gagne, 1965, p.57).

Thinking in terms of a specification of objectives, the work of Ausubel in particular offers a subtle yet relatively straightforward approach to resolving the tensions between being too specific and being too general, for achievement of the intellectual skills above the rote learning level can widely be lauded as worthwhile educational objectives.

Two basic dimensions emerge as a basis for the presentation of objectives:

● *Abilities* – That is the intellectual skills being promoted in the learner.
● *The principles, concepts and exemplars* which make up the cognitive

frameworks of subjects, and furnish the realisable levels of generality at which statements of objectives can conveniently be made.

Both provide initial guidelines for the structuring of curriculum units.

Abilities

Common elements have already been noted in the schemes of Bloom, Ausubel and Gagne. It is suggested that for broad structuring purposes a four-fold division of the abilities dimension is sufficient. To clarify again the terminology:

- *Recall* refers to the process of remembering, so that assessment of this ability involves asking questions that require recall of memorised material. The answers alone will not provide evidence of whether the memorising has been with or without understanding. In the latter case the process assessed will have been mere rote learning.
- *Comprehension* (equatable with understanding or meaningful learning) is used to cover the three sub-categories of translation, interpretation and extrapolation in Bloom's scheme, the various skills associated with Ausubel's meaningful learning and Gagne's concept and principle learning.
- *Problem solving* is extended to include Bloom's application, analysis and, to some extent, synthesis and evaluation categories, and is associated with enquiry-based learning.
- *Creativity* covers other elements of Bloom's synthesis category. The ability to 'synthesise' may be peculiarly appropriate to the traditions of geography. For the curriculum as a whole, however, a wider definition of creativity than 'synthesis' is required. Building in objectives associated with creative and spontaneous work should be a corrective to those who would define objectives too narrowly.

Principles, Concepts and Exemplars

This second dimension is clearly bound up with the first, for principles, concepts and exemplars form the raw materials of the intellectual processes grouped under the heading of 'abilities'. Principles, concepts and exemplars can all furnish educational objectives. The contention here is that it is the principles and concepts representing the higher levels of generality that provide the most convenient and cogent structures for curriculum planning. This is because they derive from the structure of knowledge itself, and they provide selecting mechanisms for the choice of content, thus avoiding proliferation. At the same time, they do not offer panaceas, and should not be deterministically or over-prescriptively applied. Such specifications can be implemented in such a way as to make the categories too clear-cut, but were in general sensitively, and differently, used by the Schools Council projects associated with geography.

Three levels of generality are identified, and the function of each is illustrated by reference to an environmental topic: water supply.

Level A: Principles or Key Ideas

As previously noted, these are formed by the linking of concepts. In Peters' definition, they are higher level assumptions or rules 'that can be appealed to in order to substantiate and give unity to lower order ones... Evidence that the principle has been grasped is provided if a person knows how to go on and deal with new situations in the light of it' (Peters, 1967, pp.18–19). The following principles or key ideas could thus be chosen as a basis for a unit on **water supply**. While reading this list, identify the concepts built into each of these statements of principles:

1. The supply of water is associated with the natural hydrological cycle.
2. Essential to life, water supply has always been a critical factor in the siting of settlements.
3. Water supplies do not necessarily occur where they are most needed.
4. The growth of population and large urban complexes has increased the need for extra-local water resources.
5. Technological and economic factors strongly influence the provision of water supplies in a situation of growing industrial and domestic demand.
6. Various hazards may constrain the provision of a reliable water supply, including environmental (drought), economic (cost escalation) and political (border disputes).
7. While on the face of it water supply is an inexhaustible resource, in practice cost factors make it a scarce resource.
8. Water supplies are prone to pollution, and such supplies may constitute a health hazard.
9. Water supply can be viewed as an extractive industry, affecting amenity, particularly in rural areas.
10. In some areas water supply is one of a number of competing land uses, with consequent conflicts of interest.

While at first sight these generalisations may seem to be citations of the obvious, they do provide a check-list of ideas which form the basis of worthwhile objectives, in turn assisting in the selection of significant content, and inhibiting the choice of the trivial or peripheral.

Level B: Concepts

In each of the above statements, more detailed concepts need to be identified. In the last statement, for example, apart from 'water supply' itself, 'competing land uses' and 'conflicts of interest' are manifestly important concepts. Yet these are highly abstract. How do they relate to such terms as 'river' or 'reservoir', which must also be classified as concepts? A breakdown of the general term 'concept' is needed, if only to draw attention to some entrenched differences of usage. Two dimensions are explored: the abstract–concrete and the technical–vernacular, to give a four-fold classification (Figure 5.1).

		Dimension 2	
		Technical	Vernacular
DIMENSION 1	Abstract	B1	B2
	Concrete	B3	B4

Figure 5.1 Dimensions: abstract, concrete, technical and vernacular concepts

The **abstract–concrete** dimension has important connotations in relation to the development of children's thinking (*see* Chapter 6), but is introduced here to distinguish roughly those phenomena which belong to the world of reasoning from those which belong to the world of direct experience. The **technical–vernacular** can be regarded as synonymous with the **scientific–everyday (or vernacular)** concepts of Vygotsky (1934). The first group (technical/scientific) covers those concepts decisively influenced by teaching; the second (vernacular/spontaneous) those concepts likely to be developed without systematic teaching. The distinction is not a black and white one; the term 'market', for example, can rank as a concrete concept in the vernacular sense, whereas its technical use in economics is abstract. But so long as the distinctions are not reified, that can serve a helpful practical function. Pursuing the water supply illustration, the following is one possible exemplification of the four types of concepts.

Level B1: Abstract Technical Concepts

- Physical concepts, e.g. the hydrological cycle (and its breakdown).
- Change through time, e.g. sequent occupance, technological change.
- Distribution and access, e.g. friction of distance, cost distance.

- Economic concepts, e.g. demand/supply, competition (for land).
- Resource, e.g. resource management (link with decision making).
- Amenity, e.g. pollution and conservation.
- Hazards, e.g. environmental, economic, political.

The connection between abstract technical concepts (B1) and principles (A) is close. Abstract technical concepts combine with each other and with more concrete concepts to form principles. To repeat: these abstract technical concepts and/or principles provide about the right level of generality at which to state objectives, for at this level the objectives do not over-proliferate; nor, on the other hand, are they too vague.

Level B2: Abstract Vernacular Concepts or Key Concepts

Levels A and B1 provide some but not all the checks needed to ensure that the objectives selected are worthwhile. Abstract vernacular concepts have been referred to as *key concepts* on both sides of the Atlantic. Examples of them can be found in the Schools Council Project *History, Geography and Social Science, 8-13* as shown below. Their relevance to the water supply topic is indicated in brackets:

1. **Communication** (questions of distance and access).
2. **Power** (who is the arbiter of choice of dam sites, for example?).
3. **Values and beliefs** (attitudes of people affected by such choice).
4. **Conflict/consensus** (between groups engaged in the issue of site choice).
5. **Continuity/change** (growth in demand for water, technological change).
6. **Similarity/difference** (comparisons/contrasts in time and space).
7. **Causes and consequences** (demand for water supply, environmental factors in choice of supply area) (Blyth *et al.*, 1976, p.94).

These key concepts are no doubt less directly helpful in the detailed structuring process, as they overlap considerably. They are essentially trans-disciplinary, and will figure in virtually all social science schemes. They have been used in world studies schemes in the field of curriculum integration, and in history schemes in subject-centred timetables. They have unfortunately been less used in geography schemes, though they have important potential there too. Without such criteria of choice, the selection of content may become over-narrowly subject-preoccupied. They bring into focus the notion that geography, among other things, is part of the social science complex, and should be issues-permeated and problem-orientated. They are, therefore, among the arbiters of 'worthwhileness', and can be seen as some of the 'uniting concepts' of human experience, with a universality which gives them a 'central importance' in social education.

Level B3: Concrete Technical Concepts

In the water supply example, the following would represent concepts of this type:

- Spring/well/dam/reservoir.
- Porous rock/aquifer/ground water.
- Pipeline/aqueduct/purification plant.
- Gradient/'head' of water/regulation of flow.

Many children will probably pick up some of these concepts without systematic instruction which would, theoretically, put them in the 'vernacular' category. It would, however, be advisable for the teacher not to make such assumptions, but to include them formally in the course unit. Even then, care must be taken to see that these concepts are not acquired purely at the level of verbal definition; indeed, this cannot be counted as concept acquisition. Thus to claim meaningful learning had been achieved, the child would have to show the ability to distinguish between the attributes of, say, a spring and a well, perhaps by labelled drawings.

Level B4: Concrete Vernacular Concepts

These would include, for example, rain, cloud, river, lake, hill, upland, rock, soil, sea, town, sewage, health, sanitation and so on. We would probably, though perhaps perilously, assume that children, by the secondary stage at least, had acquired an understanding of such concepts in the normal course of events. The concrete concepts have a distinctive role to fulfil in structuring the curriculum. They are the raw material for the more composite concepts and principles at levels A and B1/2. In return, these more general levels help to make explicit the relationships between the concrete concepts, as in the case of the hydrological cycle.

Level C: Exemplars

The term 'exemplar', though perhaps clumsy, is chosen in preference to 'fact' because:

- 'facts' have a strongly verbal connotation, and we should be thinking in terms of more than just verbal data;
- 'facts' can easily be taken to include 'concrete concepts' as well; and
- the term 'fact' has in geography acquired a rather soured image.

This is a pity, for geographical facts should be seen as fascinating. Properly selected, they are components of each child's cultural property or what one American writer has referred to, as 'cultural literacy' (Hirsch, 1987). It is they which, in individual place contexts, add reality, flavour, and vigour to lessons at the same time as the concepts and principles are providing, it is hoped, a structure and the prospect of intellectual rigour. It is a great pity that right-wing thinking so crudely and stridently claims that the acquisition of facts is the true test of an educated person, and it has to be said that Hirsch's specification tends towards this standpoint. Those who do not know the length of the Nile are not really geographers! The critical point is that exemplars should be:

● Treated as resources – presented not only as verbal data, but also in photographic, cartographic, diagrammatic, tabular and numerical forms.
● At the same time always regarded as means to ends, and not as ends in themselves.

Each 'level of generality' has thus a distinctive part to play in what might be termed the orchestration of curriculum planning. The levels provide for the teacher a conceptual framework which, together with a consideration of the different abilities to be assessed, offers a firm foundation for the selection of the content, materials and learning activities which make up the curriculum unit.

As has already been mentioned, the use of the two dimensions of 'abilities' and 'principles, concepts and exemplars' provides guidelines for the initial structuring of content. Consideration of these dimensions alone, however, could lead to an intellectualising of the curriculum which does not match up to our previously accepted aims (*see* Chapter 1). It is vital, in addition, to regard the development of attitudes and values as an aim of fundamental importance. This key dimension obviously has further consequences for the selection of content, and will be considered in Chapter 9. Geography as a subject stands squarely at a bridging point between these intellectual and social dimensions, not only as all subjects do, but also because so much of its substance derives from the social and environmental contexts in which people are placed.

Thus the following set of **key ideas** illustrates, say for GCSE level, an intellectually rigorous and socially responsible formulation of key ideas or principles as a basis for the study of recreation and tourism. It provides in effect a statement of objectives which can be related (through, as previously indicated, the mediation of place) specifically to geography, or alternatively to more integrated cross-curricular schemes in the humanities or social sciences. As previously indicated, each statement of principle needs to be broken down, as appropriate, into component concepts. These are not followed up below.

Key Ideas or Principles as a Basis for the Study of Recreation and Tourism

1. The presence of attractive coastal, mountain and other types of scenery provides a setting for physical and mental refreshment for people during their leisure time. The need for such refreshment, and

the degree of affluence required to meet this need, is most characteristic in advanced industrial societies.

2. In general, holiday-makers tend to move from:
    (a) inland areas to the coast;
    (b) lowland areas to the hills and mountains;
    (c) urban areas to rural areas (or from less to more attractive urban areas – resorts);
    (d) cooler to warmer areas;
    (e) wetter to drier areas.

3. Such movements have been facilitated by:
    (a) improved access and increased mobility resulting from technological change: foot → stage-coach → railway → automobile → aeroplane;
    (b) increased affluence and amount of leisure time.

4. Technological change has also affected the balance of competition between resorts, e.g.:
    (a) nineteenth-century resort growth where rail links were established;
    (b) twentieth century:
        (i)   resort growth through motor contact;
        (ii) resort growth through air contact;
        (iii) switch from mass holidaying at home, to package tours abroad, made possible by improved communications (travel agencies, telephones, computerised links, etc., and stimulated and programmed by media advertising).

5. Tourist areas compete for custom on the basis of:
    (a) scenic attractions
    (b) climatic attributes
    (c) quality of access.

6. Their success also reflects human decision making in terms of tourist flows:
    (a) provision of facilities, such as hotels and entertainment, designed to attract a particular clientele;
    (b) subjective perceptions and preferences of consumers (influenced by the nature of the facilities provided, and by advertising through the mass media).

7. The tourist industry may form a vital component in the national resource base. Consequently, the influence of governmental decision making and intervention may be crucial.

8. This resource base has the disadvantage in many cases of being subject to cyclical fluctuations:
    (a) seasonal, with different weather conditions concentrating the demand on particular seasons;

(b) long-term financial cycles (reduction of money spent on holidays representing a possible means of cutting costs in times of financial stringency);

(c) changing periodic consumer preferences.

9. Partly because of this seasonal basis, many tourist resorts have other urban functions, often as residential areas for commuters or retired people, or as high-class shopping centres.

10. The tourist industry exemplifies space–time tensions, being the seasonal equivalent of the diurnal 'rush-hour' in large towns, with associated traffic and pedestrian congestion.

11. The tourist industry in some cases carries with it particularly heavy environmental costs, in the degradation of coastal, mountain, moorland and forest areas, by campers, hikers, skiers and motorists.

12. While the tourist industry is a significant wealth earner in some developing countries, it should not be assumed that the benefits necessarily outweigh the drawbacks.

Such an outline highlights the range of objectives to be pursued.

It draws in varying degrees on all geography's traditions and, if pursued through place studies at different scales, can be claimed to be distinctively geographical. These frameworks of principles and concepts, as we have seen, have the further advantage of making it more likely that the overall content chosen is balanced and of intellectual and social worth. A tension which has to be resolved is that statements of objectives can too easily be linked solely to the specific and the objective, rather than the more general and subjective. The schemes suggested above take this tension on board, and stress that the values and attitudes dimension is just as important in geographical education, and vice versa. They contribute to the balance between geographical, educational and social variables which is the central thrust of this text.

Curriculum Frameworks

Over the years, curriculum planning processes have been linked with models or frameworks for planning, outlining the variables and detailing the sequence of events taking place. One of these which acquired both support and criticisms was Wheeler's 'simple curriculum process' (Figure 5.2).

This certainly picked out the key variables, but such linear sequences must misrepresent the subtleties and tensions of curriculum planning. Thus while the specification of aims may be an obvious prelude to curriculum planning, and final evaluation might provide the feedback to legitimate continuity or change in the aims statements, there needs also to be feedback, feed-forward and evaluation going on concurrently. The CERI/OECD (1972) scheme (adapted by Marsden, 1976) (Figure 5.3), while identifying

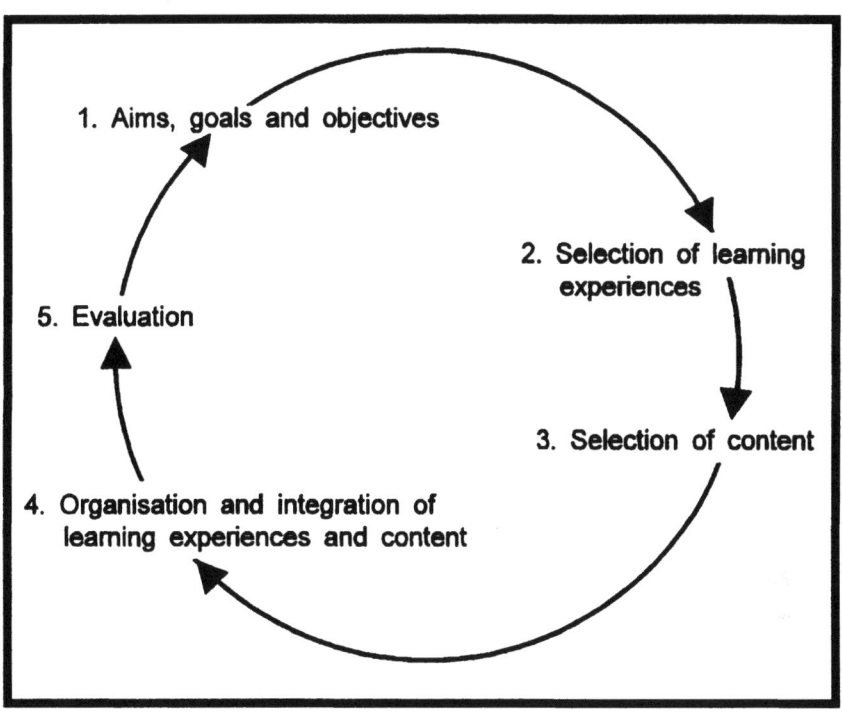

Figure 5.2 Wheeler's 'simple curriculum process'

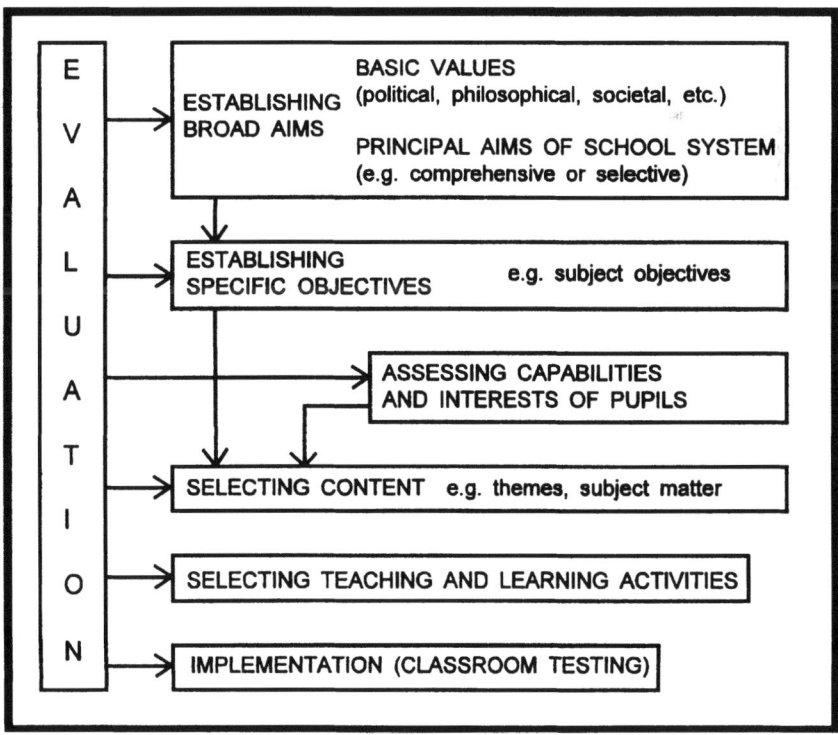

Figure 5.3 Evaluation in curriculum development

similar variables, makes more explicit the ideas that:

(a) values have to be considered as part of the context;

(b) the evaluation process needs to be permeate the whole and not just be bolted on at the end;

(c) assessment and testing should be integral to the process of learning;

(d) the subject (or cross-curricular) dimension must be part of the planning, to ensure authentic content.

A useful example of a diagrammatic framework which suggests no particular ordering but identifies the different ability variables present in both cognitive and attitudinal (affective) dimensions is shown in Figure 5.4, relating eight general objectives and modes of scientific enquiry (Schools Council, 1972, in Pring, 1989, p.40). It helpfully attaches to the centre the links between enquiry learning and the processes that are presented as scientific, though are common to other disciplines. It is interesting to translate the categories as laid out here into a geographical mould. Some of course would be replicated, or certainly overlap.

Simple models such as these can be offered in many guises and, in themselves, are of perhaps limited benefit. At they same time they do suggest that in the curriculum making process, prior thinking and planning is crucial, and involves a balanced meshing together of a number of variables in some kind of sequence.

Guidelines

While it is necessary to divide the curriculum process into a sequence, it is also something that needs to be seen whole, for the basic ground rules are of little moment if they cannot be translated into effective teaching schemes and learning activities, the focus of the next chapter. In linking this chapter with the next, it is useful to think in terms of a set of guidelines, in effect a synthesis of principles and practice, presented as a set of *Educationally Worthwhile Activities* (based on ideas of Raths, 1967).

These are offered as a check-list of criteria for what is educational worthwhile. They do not follow a strict order. An activity is likely to be educationally worthwhile if it:

- involves pupils in **meaningful learning** based on conceptual frameworks drawn, as appropriate, from the disciplines of knowledge;
- asks pupils to **examine in new settings** ideas previously studied as a means of reinforcing understandings;
- requires pupils to rehearse, rewrite and **refine** their initial efforts;
- asks pupils to engage in **enquiry** with ideas, applications, and current social and environmental problems;
- permits pupils to make **informed and balanced choices** in carrying out the activity and to reflect on the consequences of their choices;

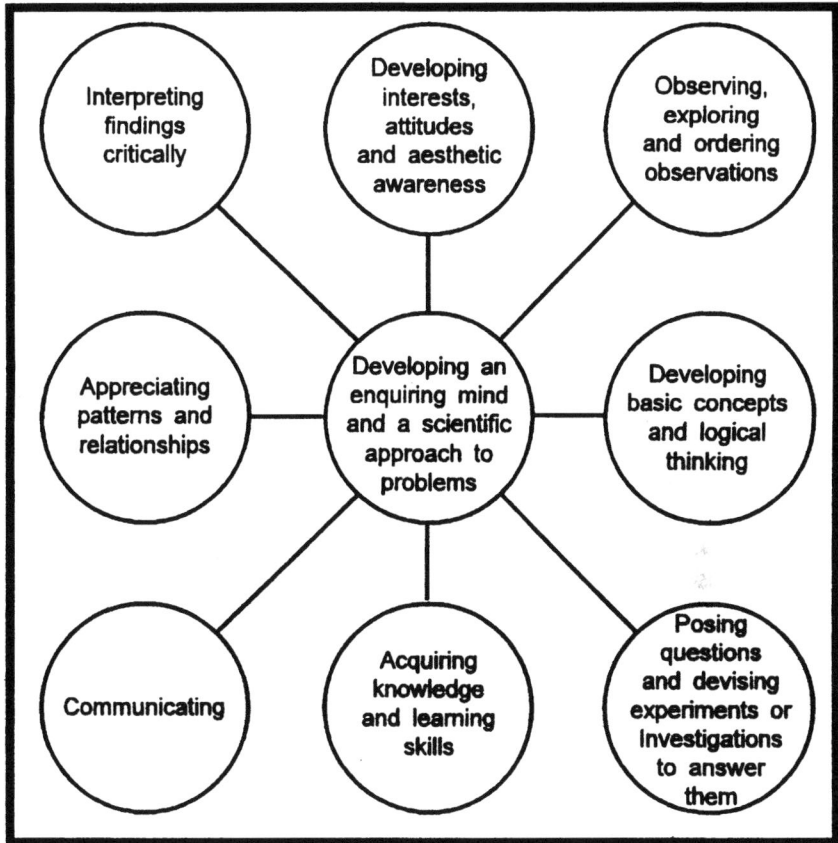

Figure 5.4 Objectives and scientific enquiry

- assigns to pupils **active** rather than passive roles in learning;
- develops an **intrinsic interest** among pupils in the topic(s) chosen for study;
- promotes **constructive attitudes and values among** children in relation to the issues being covered;
- introduces children to **content** that is up-to-date, accurate and free from bias and prejudice;
- can be accomplished successfully by pupils of different ability levels, by **matching** the teaching methods and learning strategies to individual needs.

These guidelines take us beyond the initial specification of objectives, however. They shift the discussion from planning, structures, and thoughts about outcomes, to classroom activities. They set up a whole series of questions about the relationship between curriculum objectives and the processes of teaching and learning, which will be pursued in the next chapter.

CHAPTER 6

Teaching and Learning Geography

The Development of Spatial Cognition

In the intellectual growth of children and young people, checking the development of **spatial cognition**, particularly in map reading, is a central concern for the geography teacher. A number of theories are germane to this important aspect of human learning. So far as human territory is concerned there are, for example, in the earlier stages of child development, moves from:

- **action in space,** to
- **perceptions of space,** to
- **conceptions about space.**

In more general psychological terms, the process in turn reflects the shift from egocentric, simple and undifferentiated views of the world, to those more complex, differentiated and distanced from the individual. By the secondary phase, the hope is that pupils will have reached the third stage of development, conceptions about space. But as always, we must recognise overlaps between the categories and cannot assume they are achieved consecutively. Each gives support to the others at different ages: thus the physical activity of field work and use of the senses are important contributors to the development of intellectual conceptions. Though this book is directed at methodologies underpinning the teaching of older children, it is important that secondary teachers understand the developmental sequence, for at the age of transfer from the primary phase, there are huge differences between levels of understanding in this area. Spatial concepts are associated with notions of:

- **distance**
- **direction**, and
- **relationships between objects in space.**

In their early organisation of space, children acquire topological rather than Euclidean or projective understanding (Rhys, 1972). Hence the first spatial properties which can be represented are to do with proximity, spatial succession, and surrounding or enclosure, rather than with the continuity of lines or absolute distances. A useful way of making these ideas more concrete is to think of the simplified properties of the Intercity or London underground topological maps, as compared with the complexity and diversionary 'noise' of maps accurately represented in Euclidean terms, in which concepts of distance and direction are strictly preserved.

Following the egocentric stage, an intermediate stage of fixed and concrete systems of reference is needed as a means of orientation to larger scale environments, even local ones. Co-ordination is by reference to familiar landmarks such as the room, the home, the school, and the route to school. Thus children can deal better with route maps, mentally tracing their locomotion through an area more confidently than with overall survey maps, like Ordnance Survey maps, which are representations of a general configuration of objects. The 'local' scheme of things has a definite relationship with the child's known and experienced world. This early stage of being tied to the parochial is known as **domicentricity** (Hart and Moore, 1973). A sense of familiarity with the home area is potentially helpful as a basis for moving outwards. Maps and photographs can be orientated by finding home first.

The process of moving out from this personal territory is known as **decentration.** An unfortunate outcome of the face-value interpretation and application of Piaget's studies, which laid some of the groundwork for these ideas, was that it tended to promote the deficit view that younger children needed to be tied to the familiar for a long period of time, even the whole of the primary phase of schooling (Martland, 1994, p.37). Many subsequent studies have suggested, however, that this promotes under-expectation, and decentration should be encouraged as early as possible. Later work has been critical both of Piaget's methodology and its subsequent inappropriate application to work on the development of mapping skills (Spencer et al., 1989, p.133). Furthermore, environmental learning is now judged to be not dependent on the development of spatial cognition alone (Matthews, 1992, p.83). Thus Matthews, and Spencer and his colleagues, have emphasised the importance of not becoming preoccupied with a narrow mathematical and linear approach to the development of mapping skills, but of placing these in the broader context of the development of environmental experience and understanding.

This broad field of spatial cognition forms a major part of what has been referred to as the most geographical form of communication, namely **graphicacy**, which covers not only the development of map reading skills, but also the interpretation of photographs and other forms of graphic communication (Boardman, 1983).

Forming concepts in geography

Early stages

Meaningful learning is linked with the acquisition of concepts (*see* Chapter 5). A major group of concepts encountered in the early stages of geography teaching are those **concrete vernacular** concepts which relate, for example, to physical features of the landscape, such as 'river', 'valley', 'hill', 'mountain' and so forth. In the development of a concept such as 'mountain', the first step involves appropriately associating the **criterial attributes** which make up this concept, which are shown on Figure 6.1.

How do we know whether or not a child has grasped the concept of a mountain? Let us explore the theory through another concrete everyday concept, that of a chair. If a child comes into a room and we ask that child to point out a chair, and this is done accurately, would that be sufficient evidence that the concept has been acquired? The answer is surely no, especially if the room and its furniture are familiar, for what might have

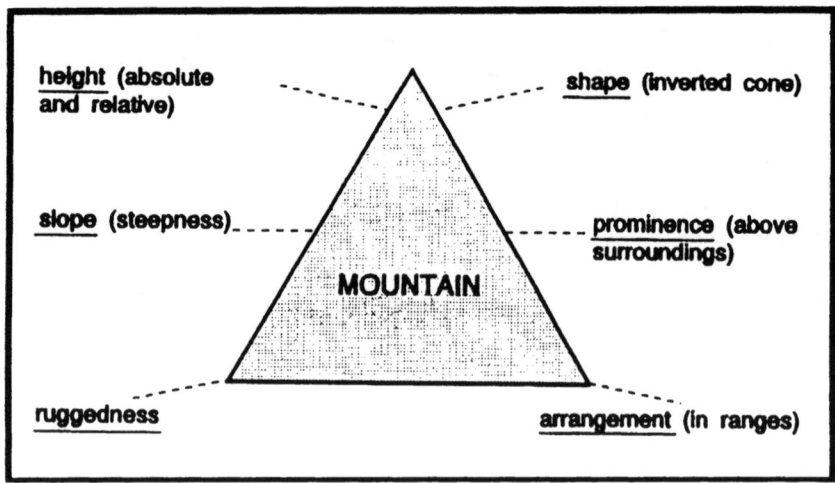

Figure 6.1 Mountains: basic attributes (concepts)

been achieved is mere recognition. What would be more satisfactory evidence would be if the child could:

- make a labelled drawing of a chair; or
- distinguish a chair from a table (which has many of the same attributes); and
- make clear the differences (in the criterial attributes).

Returning to the case of the mountain, neither would it be sufficient for the child merely to repeat facts learned about mountains. We need more evidence than this to be in a position to accept that an adequate level of conceptualisation has been achieved. Meaningful learning requires something more than the ability to verbalise the word 'mountain', or even state that 'Everest is a mountain' or that 'the Himalayas are a range of mountains'. Evidence is needed that the child can identify exemplars and non-exemplars of particular criteria; for example, distinguishing mountains successfully from hills, on diagrams, in photographs or in the field (*see* Chapter 7).

It is also interesting and valuable to build discussions on children's own ideas, including what may appear as misconceptions. They can offer unusual insights. On enquiring into children's ability to distinguish a mountain and a hill, for example, one Year 7 teacher found, in addition to the more obvious distinctions, unusual ideas like: 'you can get up a hill in less than three hours, but not up a mountain'; or 'you can go up a hill in a car but not up a mountain'. Such 'alternative' views also usefully suggest that even at secondary level it is important to link the bald concepts of physical geography with the human experience.

The basic concept of 'mountain' should not be difficult to establish, given proper stimulus materials. It is the task of the geography teacher to enrich the concept, once satisfactorily established, perhaps by pursuing distinctions or contrasts between different types of mountains in terms of appearance, human occupancy and origin. It is indeed psychologically appropriate that the National Curriculum stresses the idea of linking the themes of geography.

Figure 6.2 connects the progressive development of the concept of mountains with Bruner's idea of a **spiral curriculum**, which involves revisiting, reinforcing and refining (but not mere repetition) of concepts at successive stages. Such is the essence of the key educational idea of **progression**, which can be thought of in geography in terms of moving from:

- the familiar to the unfamiliar;
- the near to the more distant;
- the concrete to the abstract;
- the smaller to the larger scale;
- the simple to the more complex in terms of:
  – breadth of coverage;
  – depth of coverage;
- a more to a less limited range of skills.

Progression is thus a longitudinal concept. The associated horizontal one is **match**, in which, as we have seen, materials are differentiated so as to meet the needs of the individual pupil at any particular stage. One time-honoured way of addressing this issue is to begin with core activities, thereafter providing **enrichment** materials for the more able and **reinforcement** material for the less able. Information is therefore needed on whether:

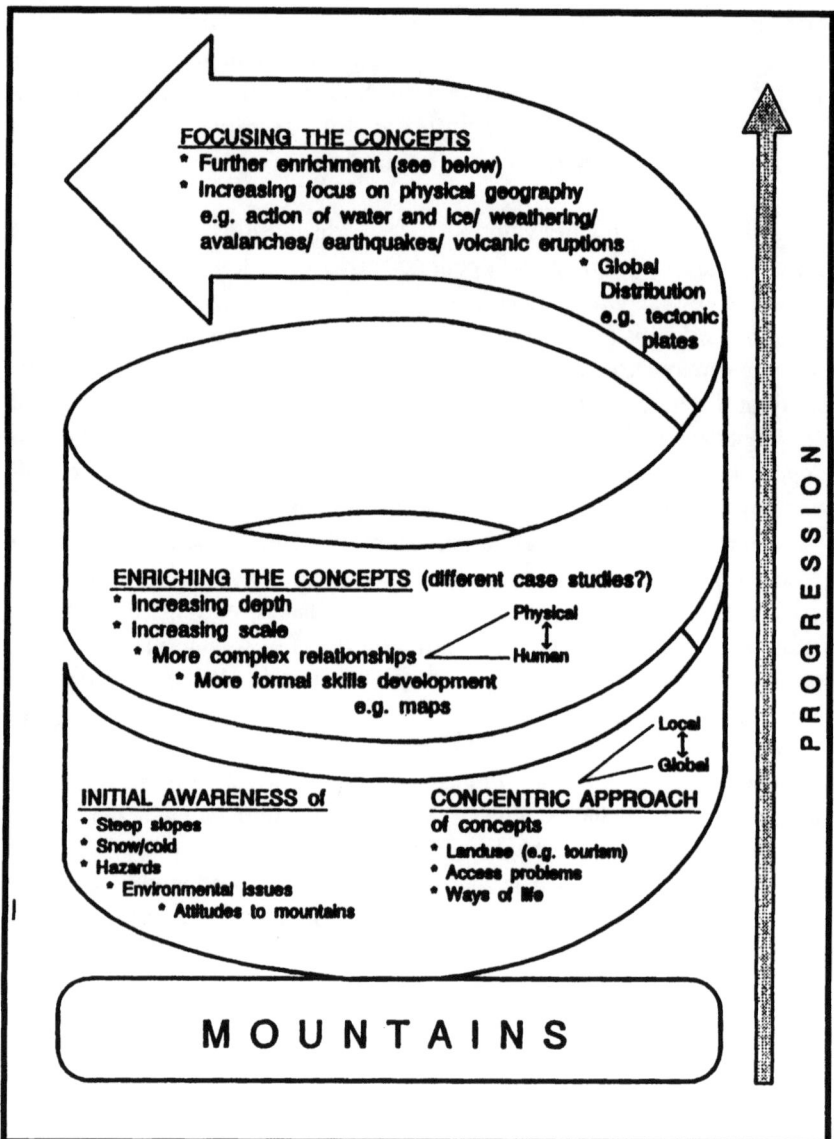

Figure 6.2 A spiral curriculum: primary geography

- the pupil is showing herself or himself able to apply the skills and ideas which the activity demands;
- success is out of reach with the current set of materials and strategies;
- success might be achieved by changing the approach and way of thinking;
- a new challenge is needed to extend skills and ideas;
- interest and perseverance is being shown in the activity;
- the classroom and social context in which the activity is being pursued is appropriate.

On the basis of this evidence, decisions have to be taken whether simplified reinforcement activities or more complex enrichment ones are needed; whether it is a case of needing more encouragement and confidence-boosting, or whether the whole activity is misconceived and a new one needs organising.

At the secondary school stage, we are concerned with the transition from and interaction between what, in Piagetian terms, are referred to as the **concrete and formal operations stages**, and increasingly with the latter. Rhys (1972) has argued that during the years 12–15, pupils are increasingly able to engage in thought that goes beyond practical problems and concrete situations, since they can now make use of abstraction and generation, and can employ hypothetical and deductive reasoning to deal with problems and possibilities not encountered before (p.99).

These outcomes will only happen for many pupils where there is sensitive and knowledgeable teaching support. Slater (1982), for example, pointed to their difficulties in thinking abstractly about their environment, even though in Piagetian terms at the formal operations stage of development. The root of the problem seemed to lie in the fact that underlying geographical concepts were not being taught or learned. Pupils do not emerge with conceptual understanding from an engagement involving learning masses of content alone. Exercises requiring application, problem solving, simulation and role playing are all recommended, reflecting Ausubel's ideas (*see* Chapter 5).

As we have noted, a problem of Piagetian-type research is that it can lead practitioners to believe that because some pupils, for different reasons, cannot achieve more complex tasks, this is a pre-determined characteristic. It too easily leads to under-expectation of pupils. The school, as Vygotsky (1934) has urged, is the responsible body for presenting demanding, problem-orientated tasks to the student. With assistance every child can do more than s(he) can by herself or himself, though only within the limits set by the state of cognitive development. But 'the only good kind of instruction is that which marches ahead of development and leads it' (pp.103–4).

Language

While geography makes appropriate claims to be a subject which stresses the cartographic and the visual, this is not to say that written and spoken language is not equally crucial to this subject as to others. Milburn (1972) made the critical point that the formation of geographical concepts is closely associated with the development of an appropriate vocabulary (p.107). One problem is that repeating the correct vocabulary, which can be achieved by rote learning, does not mean that the concepts have been meaningfully acquired. The acquisition of concepts and an actively associated vocabulary are complementary aspects of intellectual development. Accurate verbal expression also helps to consolidate and refine meanings (p. 119). Confidence in handling and deploying the vocabulary also assists success and consequently positively motivates.

The language of the classroom can be divided into teacher talk and pupil talk, and into teacher writing and pupil writing. All of these are important (Barnes, 1969). Teacher talk and pupil talk are clearly central to Socratic-type dialogue in the classroom, to be considered later under teacher questioning. Here we will concentrate on teacher writing. In a subject like geography, there are many traps for the unwary. The following issues need particular attention.

Readability

An important element in teacher's writing, or that in textbooks, is the readability level, readability referring essentially to the degree of ease with which the material can be read by a group of learners (Williams, 1981, p.71). The appraisal can be based on liveliness of style and the interest the material generates, but the more fundamental issues are of complexity of grammatical structures, length of words, density of text, abstraction of concepts, the technical language used, font size, design and other features. The Flesch formula is a well-known test of readability, and is based on syllable counts and sentence lengths (*see* Boardman, 1982).

Technical and vernacular language

The key problems lie in the relationships between technical and vernacular language. The former often involves the translation down of the impersonal and condensed language of academia. Here the ways of saying characteristically tend to 'keep their distance and shift discourse from its specific context to a very general one, from the world of you and me and him to the world of "one" and the passive voice' (Rosen, 1969, p.132). Rosen has pointed out how teachers have been critical of lively and involved pupil language in their writings because it is more emotive and personalised than the formal rule books allow, and less than accurately expressed (Rosen, 1967, p.119).

Geikie was aware of the problems over 100 years ago, and urged that the language first used, whether written or through talk, should be 'homely and conversational', and that class-books should be avoided in the early stages (1887, p.12). In his analysis of textbook language, Rosen (1967, pp.120–1) highlighted the difficulties found in the pre-packaged academic discourse of a well-known series entitled *Geography for Today*, in which the packed association of terms like 'aridity', 'temperature ranges', 'rain-carrying winds', 'rainfall distribution', while each carrying important meanings, denied the likelihood of a response based on genuine pupil comprehension, still less of arousing interest. At the same time, technical generalising language is a vital basis of advancing thought, and ordering and systematising experience. It has to be worked towards and used, but more student-centredly than has often been the case. Even in the more attractive presentations of today's textbooks and multi-media packs, there is a need for vigilance in identifying the many problems present in trying to make geographical concepts meaningful.

Sub-systems of linguistic conventions

Further, most academic disciplines have their own sub-systems of linguistic conventions and technical language, some of this readily labelled 'jargon'. There are differences in the language of a science, or geography, or history, or English literature, or a social science. In geography there is the additional issue of place-related concepts, which mean different things in different places (Milburn, 1972, pp.118–9). The language of geography also contains the specific vocabulary of foreign place names, often difficult to pronounce and to spell.

Homonyms

There is in addition the related problem of homonyms, words often used in the vernacular but which have acquired a technical meaning in geography: some specific to geography/geology, like 'joints' and 'faults'; others used in a range of subjects, like 'market', or 'energy'. To pupils, they can be a major source of ambiguity. Geography draws on a wide range of concepts from different fields of study, particularly the social and earth sciences, as well as from common-sense language. The subject is less divorced from reality than other subjects, and may seem to contain less technical language than others. In fact the difficulties are only rather more hidden.

Factors influencing Progression

Ability Differences

Whether geography as a subject taps particular intellectual abilities or a special 'intelligence' appears still to be an open question. As already noted, the particular ability which impinges most characteristically on the content and skills of geography is spatial cognition. There are undoubted differences in the pace and depth at which pupils acquire spatial concepts. A major problem is that there appears to be a maximum difference in levels achieved at the transition between primary and secondary phases. For secondary teachers, the situation is acute, because it is hard to distinguish failure or success on the basis of children's abilities alone, in that their experiences of spatial problems at the primary phase are so disparate (though hopefully less so with the coming of the National Curriculum).

Special Needs

The vital response to the perennial problem of how to deal with differences in ability must be found in strategies that address questions of match and progression. How to deal with truly low-ability children is a specialist subject but, again, one that has to be kept in mind in a situation when many 'special' children are now taught in mainstream schools. Traditionally, special needs children, like primary children, and lower ability children in the secondary school, have been thought in need of protection from subject-centred curricula. This attitude has tended to change with the coming of the National Curriculum. If the rest of the school population is studying geography, then the concept of bringing special needs children into the mainstream is undermined if they are denied the subject entitlement.

Over the years the Geographical Association has produced guidance on teaching slower learning children (Boardman, 1982; Corney and Rawling, 1985), rather than those defined as having special needs, but this situation has changed more recently (Dilkes and Nicholls, 1988; Sebba, 1991, 1995). In one of its last cross-curricular guidance documents, the NCC addressed *The National Curriculum and Pupils with Severe Learning Difficulties (No. 9)* (1992) in which, perversely, advice was given relating to the core subjects, technology and cross-curricular themes, but not to the other foundation subjects.

In its final report, the National Curriculum Geography Working Group drew attention to the difficulty of drawing up Statements of Attainment and Programmes of Study appropriate to special needs children, including children unable to communicate by speech; children unable to see well enough to read maps or photographs; children with restricted mobility who had not developed normal concepts of space, and so on (1990, p.79).

The following categories all need sophisticated and dedicated attention, not least when required in the mainstream school:

1. Pupils with exceptionally severe learning difficulties including:
   – profound and multiple learning difficulties, some of whom may have associated challenging behaviours
   – exceptionally severe learning disabilities resulting from, for example, multi-sensory impairment.
2. Pupils with other learning difficulties including:
   – mild learning difficulties
   – moderate learning difficulties
   – specific learning difficulties
   – emotional and behavioural difficulties.

3. Pupils with physical or sensory impairment.

4. Exceptionally able pupils.

It is doubtful whether subject teachers are as yet in a position adequately to grapple with what are in some cases highly specialist problems, but there is some agreement that maximal use should be made of opportunities for use of:

- communications skills linked with enquiry-based learning development in the local area;
- specialised and compensatory IT support;
- other technical aids, e.g. braille and tactile maps;
- visual materials for those with specific learning problems with text, including images of more distant places;
- human and environmental geography as bases for broader personal and social education programmes;
- sensory approaches to the environment.

With special needs children the tension between stress on the familiar and taking them out into the wider world has particular relevance. Geography can at its best provide a rich resource of concrete and lively experience, but it needs teachers with particular skills and sensitivities and a broad grasp of the curriculum as a whole (*see* Hughes and Thomas, 1994).

Gender

Much evidence has been collected to suggest that gender differences in spatial ability manifest themselves at particular stages. Thus Matthews (1984, 1986) has noted that from the age of 8 boys in particular are allowed freer range than girls and, as a result, have more awareness of the environment around them, encompassing a broader territorial sweep. By the age of 11, he found girls less equipped to understand spatial relationships and more inclined to resort to simpler frames of geographical reference. Thus the differences are socially constructed rather than genetic. Matthews urges compensatory action, not least because the deficit can encourage girls to feel geography is not their scene. One obvious ploy is to limit the amount of narrow map reading work in the early stages of the secondary phase, where it is sometimes pervasive. Clearly, there is a delicate balance here, for it would be counter-productive to downplay a basic geographical skill. The vital need is to be sensitive to the issue and produce map-reading activities that are responsive to the environmental experience of girls and young women (Madge, 1994). There is increasing agreement that these differences are psycho-social rather than genetic or environmental (Gipps and Murphy,1994). The gender issue will be returned to in Chapter 7.

The Time Factor

Most teachers would agree that time is an important component in learning. Carroll went further and maintained that it was the crucial component. The 'learner will succeed in learning a given task the extent that he spends the amount of time he needs to learn the task' (1963, p.725). Time refers of course not to elapsed time, but time spent in learning. Apart from basic pupil ability and the quality of teaching, the variables for successful learning Carroll saw as directly linked with the time factor:

1. *Aptitude* – The time needed by a learner to learn a task. Children of high aptitude for a particular task can master it more quickly than those of low aptitude.

2. *Opportunity* – The time allowed for learning. Lack of time causes inefficient learning, and may incite rote learning. Opportunity is theoretically a variable under teacher control, though teachers would argue that the pressure of external examinations and, more recently, the first phase of National Curriculum implementation, make time very scarce. Over-inflated syllabuses are a deleterious influence in education, for 'one of the most aversive things which a school can do is not to allow sufficient time for a well-motivated child to master a given task before the next is taken up' (p.728).

3. *Perseverance* – The time the learner is willing to spend in learning, which is clearly linked with motivation.

Differences in Motivation

For meaningful learning to take place, there must be some intent to learn successfully (Ausubel and Robinson, 1969, p.53). Motivation is a complex phenomenon, and is related to the achievement of basic human needs. Leaving aside physiological needs, there are three basic needs or drives affecting children's motivation (pp.357–8):

1. *The cognitive drive* – Which has to do with curiosity, the desire to know and understand. Learning activities are seen as rewarding things in themselves. Motivation of this type is therefore **intrinsic**.

2. *The achievement drive* – Which is linked with ambition, the competitive spirit, the earning of status in terms of current prestige, and the willingness to defer present satisfaction to achieve future ends, usually career goals. It is an **extrinsic** form of motivation, which may promote purely an instrumental view and an instructional approach to the educational process, seen merely as a preparation for adult life. The achievement drive is closely associated with fear of failure, which can range from extreme neuroticism to a manageable concern. In the performance of straightforward tasks, mild anxiety is regarded as a

positive motivational force. But where learning tasks are difficult, anxiety impairs discrimination and performance deteriorates.

The single-minded pursuit of examination success can have crippling academic consequences in that in many subjects, of which geography is a prime example, it has traditionally been achievable through rote learning strategies. This can result in the creation of the 'articulate idiot' defined by Bruner (1960) as 'the student who is full of seemingly appropriate words but has no matching ability to use the ideas for which the words presumably stand' (p.55).

3. *The affiliation drive* – Which is associated with the desire to please. The growing child interacts with parents, teachers and friends. S(he) has a need to please one or more groups – which one varying with different stages of physiological and emotional development. In adolescence the desire to affiliate with the peer-group becomes stronger, and in some cases over-powering. While the desire to 'please teacher' may be weaker than in the primary school, alienation is far from inevitable at the secondary stage. Possible gender differences need to be considered. Teacher influence remains, however, critically important.

The Influence of the Teacher

Teacher–pupil Relationships and Classroom Management

While this book is not primarily addressed to problems of discipline and classroom management, these obviously impinge on the success or otherwise of geography teaching. A number of generalisations, even truisms, are widely accepted. One is that pupil alienation is more likely with the authoritarian teacher on the one hand and the weak teacher on the other, than with one who is generally calm, cheerful and sympathetic, fair, more inclined to praise than blame, and intent on running without fuss an orderly class and creating a 'learning environment'. It is generally agreed that extremes of rigid authoritarianism and permissiveness should be avoided. Over-strict discipline arouses sullen and fearful reactions, and inhibits response in enquiry-based learning and class discussion. It is the death of intrinsic motivation. Absence of discipline problems, however, is no proof that pupils are in reality involved in their work. They may be soporifically acceptant of a rigidly didactic method of teaching, geared to note taking and rote learning.

At the same time, a riotous classroom atmosphere is totally disruptive of learning. While there are members of staff whose control is weak even in the most placid of suburban or rural schools, discipline problems are generally seen as concentrated in the lower streams of schools where the range of ability in the intake is great, especially in inner city areas. In such schools, breakdown of relations between teachers and pupils is widely publicised. Some years ago, Keddie (1971) highlighted as an important contributing factor to such breakdown as **low teacher expectation**. She maintained that teachers developed negatively stereotyped ideas about certain classes and certain pupils. Even though the vocabulary outside the classroom of some teachers might have seemed a progressive one, their attitude towards different classes varied, with a 'reciprocity of perspective' between teachers and 'A' stream-type pupils, while 'C' stream-type were labelled deviant and unteachable – whether or not the school was streamed. The phenomenon was referred to as the 'sociological fallacy', and subsequent research has repeatedly confirmed that school influences are important, in deprived areas more than most. OFSTED reports more recently have pointed to high- and low-achieving schools in both well-to-do and deprived social areas.

Teacher expectations communicate themselves to pupils through all sorts of cues the teacher gives out, and produce the phenomenon of the **self-fulfilling prophecy**. If teachers expect poor pupil performance they get it. Wiseman (1973) regarded the identification of teacher expectation as a major variable affecting the performance of children: 'the single most significant outcome of educational research' in recent times (p.90). It has to be said that some right-wing and other commentators have been sceptical of these arguments. Thus Eysenck (1973) denounced the 'self-fulfilling prophecy' argument as 'an educational myth' if it was suggest-ing that IQs could be changed in the direction expected of them.

Teaching Styles and Classroom Management

In achieving a balance between democracy and structure, between a libertarian outlook and good practical management, a relevant concept may be that of **pedagogic control**, advanced by Holly (1973, p.145). The idea of control is here not used in the sense of discipline, but refers to decisions taken by the teacher in relation to the organisation of learning. The direction exercised by the teacher is of materials, learning resources and framework rather than of the conduct of pupils. The strategy is based on structured independent learning, and must be powered by cognitive and achievement, as well as affiliative drives, on the part of the pupil.

Since the 1970s we have moved on to place more emphasis on the importance of the autonomy of the teacher and on **school-centred innovation** (*see* Tolley and Orrell, 1977; Skilbeck, 1984). In years past, Schools Council and other projects found that curriculum change did not take wing in schools where there was a top-down and sometimes insensitive imposition of curriculum ideas from above or from the outside. The new conception was that teachers needed to gain **ownership** of the innovation, otherwise it was likely that there would be innovation without change. Such arguments would seem entirely logical and potentially constructive, with two caveats. In the first place, imposition from above does achieve curriculum change. Thus the success of the Avery Hill and other projects, albeit avoiding top-down imposition, would not have achieved the success they did had they not tied themselves to external examination syllabuses, an extrinsic form of motivation. It can be argued that in some senses teachers no more 'own' Avery Hill than any other syllabuses, though they may find them for a number of reasons more congenial. By the same token, the examination syllabuses were in turn changed and much improved (*see* Chapter 7).

Another problem arose in that in the post-1979 desire in the world of education to combat the dangers of centralisation, the attractive alternative of school-centred innovation was in some cases advanced uncritically (Hargreaves, 1982). Thus in a case study of teacher-centred innovation, Kirk (1988) found that to accept it incautiously as necessarily good was not justified, and that enthusiastic collective action on the part of teachers did not rule out the possibilities in practice of disagreement and internal power plays which were 'owned' by some and excluded others. Consensus in the 'average' teaching force against outside forces might readily be achieved, but at the same time there could be strongly conflicting ideological assumptions within school-centred innovation (p.463). Achieving the necessary clearly agreed aims can prove more difficult in circumstances where internal forces in the school contest ownership, than when imposed from without, as in the case of the National Curriculum.

The other major difficulty in school-centred innovation has been political: giving individual schools ownership over curricula was not on the policy agenda of a government engaged in the 1980s in tightening the ratchet over the education industry as a whole.

Individualising Learning

Individualisation of learning was advanced over 60 years ago by Courtis (1930) as 'the master trend in education' (p.52). More recently, Gagne (1967) is but one who has concluded that processing of information is an idiosyncratic procedure, and avers 'that individualised instruction represents the route to efficient learning' (p.312). The main direction of the argument in this chapter has been shifting towards this position. No other strategy seems as relevant in the context of mixed ability classes, for example. One of the main justifications of flexible learning strategies is that teachers can spend more time with individual pupils in particular need of support.

Flexible learning

An interesting partial anticipation of the principles underlying individual and flexible learning is to be found in an article by Pearce in the 1929 edition of *Geography*, in which she asked pupils themselves to check in detail marked assignments, '...providing a means of self-assessment, writing down "ways in which I can improve my assignments"' (pp.298–302). One of the important school-based innovations of the 1980s lay in the promotion of flexible learning strategies. Generally this is defined as a form of **supported self-study** involving a movement from:

- teacher- to student-led learning;
- whole class-work to group and independent learning.

Students take responsibility for their own learning, within a framework of background support, negotiating with their teachers the means to achieve agreed goals, and even setting their own goals and identifying short-, medium- and long-term targets. The students therefore characteristically interact directly with resources and with their peer group, i.e.:

- plan pathways towards a goal
- select appropriate material
- research the material to solve particular problems
- have the opportunity to cooperate with others
- present the outcomes to the peer group
- self-assess.

The units of study are often designed to be covered in short blocks of time, as it is vital in this process that students obtain regular feedback. Assessment and recording are seen as integral parts of the learning process. The students can initiate tutorials with the teacher as well as vice versa. Skilled management is vital to the success of the endeavour, whether in organising student groups, making resources available, or in auditing the learning pathways. Inevitably, a wide and perhaps idiosyncratic range of approaches emerges, and different students will be engaged on different tasks at different times. They will move about regularly within the classroom, and from classroom to library and resource centre, even outside the school. Therefore, skilled classroom management is vital to the success of the endeavour.

The flexible learning strategy has been much associated with TVEI initiatives. It attaches to itself all the favourable labelling of progressive practice, such as self-ownership, experiential learning, active learning, resource-based learning, computer-assisted learning, distance learning, and holism. One problem is that the mere reiteration of these labels is taken to legitimate the approach and sanctify it as good practice. As in the primary school, there will inevitably be a variety of qualities of response under the same umbrella, some much better than others. In general, these processes should underpin good geography practice in whatever teaching style; and many proponents of flexible learning acknowledge that it is merely an extension of existing good practice. The success of flexible learning will inevitably be subverted, for example, if the materials towards which the students are tracked are biased, out-of-date, conceptually primitive, second-hand and poorly presented. Flexible learning is no panacea, though it is sometimes presented as such.

The National Curriculum to an extent creates some problems for flexible learning strategies in terms of:

- the nature of the assessment and recording of progress
- match with National Curriculum descriptions
- coverage of National Curriculum content in geography.

Fortunately, the National Curriculum still does not dictate the pathways used towards achieving the statutory requirements, and the post-Dearing changes are intended to offer more flexibility. They will hopefully reduce the assessment millstone previously envisaged. There is, arguably, no inevitable incompatibility between National Curriculum demands and flexible learning techniques, unless of course the teacher defines the two in ideologically polarised terms. This could be said to be the reaction of the narrow craft-oriented practitioner. The true professional will be able to find ways of reconciling flexible learning approaches with the demands of the National Curriculum at Key Stage 3 (*see* Chapter 12).

Questioning Skills

In whole-class or group modes of teaching one of the most vital skills is that of questioning. Questions can of course be formal and limited, but questions are also the basis of enquiry learning. As Manson (1973) maintained:

'Teachers should be able to ask good questions. A teacher should be able to ask provocative questions which will stimulate interest and group discussion. A teacher should be able to phrase insightful questions which will promote thinking and expedite problem solving. A teacher should be able to construct valid and reliable test questions which will permit him to make an accurate assessment of learning. And a teacher should be able to pose thoughtful and clarifying questions which will help students recognise and assess their attitudes and value commitments.' (p.24)

Manson proposed a typology of questions based on the Bloom taxonomy. He took the Bloom 'abilities' as a 'process dimension', which he placed as the vertical axis in a matrix against a 'knowledge dimension' which included facts, concepts and generalisations, forming the horizontal axis (*see* Marsden, 1976a, p.149). He offered examples of the sorts of geographical questions that were appropriate to each section of the matrix. This type of specification, though potentially too limiting, is useful at least in drawing attention to the range of questions that is available, and can serve to highlight how little of this range is used, for classroom practice too readily emphasises the more straightforward closed content questions and tends to marginalise the more loaded and open-ended values questions.

Good questioning will include, as appropriate, the following attributes:

- asking questions fluently and precisely;
- gearing questions to the student's state of readiness;
- involving a wide range of students in the question–answer process;
- focusing questions on a wide range of intellectual skills, and not just on recall;
- asking probing questions;
- not accepting each answer as of equal validity, though sensitively;
- redirecting questioning to allow accurate and relevant answers to emerge;
- using open-ended as well as closed questions, so that creative thought and value judgements are invited.

The process should therefore incorporate:

- a limited number of recall questions (as means to wider ends);
- comprehension questions;
- application and problem-solving questions;
- open-ended, creative questions.

These can appropriately be linked with the Ausubel categories outlined in Chapter 5.

Stimulating Interest

It is often argued that disadvantaged pupils, and others, are alienated from school because didactic methods and a content-driven curriculum menu derived from 'high culture' are imposed upon them. A preferred strategy is held to be the stimulus of interest through relevant materials, relevant being taken to mean applicability to the child's social and environmental situation. Bearing in mind our original statement of aims, an equally tenable definition of 'relevant' would be relevant to the child's intellectual, and through this personal, development. The idea that there should be an 'interest-based curriculum' for some, that is the less able, and 'real education' for others, has long been criticised. It is an illusion in any case to think that intellectual skills and 'interests' can be separated. The political tension is that in introducing comprehensive schools, mixed ability classes, and in merging the GCE and CSE examinations, right-wing opinion has been vocal in contending that there has been a shift away from intellectual content and a consequent lowering of standards. This has continued as a heated debate in the run-up to and implementation of the National Curriculum, and is reflected also in government-sponsored moves back to selective schooling.

The promotion of interest-based learning is in practice, though not necessarily in principle, associated at the macro-level with integrated rather than subject-based schemes and, at the micro-, with the use of comic-cuts-type presentations which, as we have seen, were unfortunately legitimated in materials derived from the Avery Hill project. Thus can a deficit view be implanted.

In any case, the geography teacher of today knows, however 'socially relevant', the material is, it is no guarantee in itself of gaining interest. As Marland, a practising headteacher, wrote (1973): 'Pupils can be as resoundingly switched off by the relevant as the obscure; they can be as bored by the apparently child-centred as by the adult-dominated' (p.18). The importance of engaging pupils in geography in the exploration of major contemporary social and environmental issues, as expressed in place and space, is entirely to be supported. But it would be disingenuous to accept that this is a certain, or even the most likely, way of promoting positive motivation.

Reinforcing Success: an Alternative Strategy

The unequivocal view of Ausubel and Robinson (1969) was different. They claimed, on the basis of their research, that:

> 'the best way of motivating an unmotivated pupil is temporarily to by-pass the problem of motivation and to focus on the cognitive aspect of teaching...[and] to rely on the cognitive motivation that is developed retroactively from successful educational achievement.' (p.447)

Academic success must be seen as achievable from an early stage as a result of which the teacher 'can establish a motivational beachhead for the crucial pre-adolescent stage when alienation from school would normally reach sizeable proportions' (p.445). It must be stressed that this was not seen as a narrow concern to foster extrinsic motivation. In this first place it was to do with promoting self-esteem.

Wall (1968) too, was clear that:

> 'Success, if frequent enough, leads the pupil to set appropriate goals while continued failure leads him to aim too low (to ensure a relative success) or unrealistically high....Failure in a general context of success has a different meaning from failure which is one more of a long chain.' (p.58)

Mastery Learning

Bloom regarded the main job of the teacher as fostering successful learning, and advanced the concept of mastery learning (1968). He and colleagues suggested that 'given sufficient time and appropriate types of help, 95% of students can learn a subject with a high degree of mastery', and affirmed that, leaving out the top 5% and bottom 5% of the school population, the ratio of time required as between the slow and the quick learner was 6 to 1. They claimed, tentatively, that with effective teaching this could be reduced to 3 to 1 (1971, p.51). For internal assessment purposes, they regarded the normal curve of distribution as a subversive influence. If results equated with this curve, they were to be seen as symptomatic of unsuccessful teaching, with large numbers of children inevitably viewing themselves as having failed. Bloom recognised the difficulty of applying this concept in a realistic way in the secondary school, for by this stage many children would already have experienced a long string of failures.

It may be, therefore, that in the first stages of the secondary school it would be expedient quietly to manipulate standards in the short term in order to avoid the debilitating effects of early failure. This is not to advocate any long-term lowering of standards, but to provide the springboard which might lead to an overall raising of standards through the stimulus of cumulative success, leading on to intrinsic motivation. The problems of pursuing this approach in an unsympathetic political climate are of course recognised. Two key strategies in this process are the effective use of feedback and, as we have seen, the individualisation of learning.

Feedback

The concept of reinforcing success is inextricably linked with regular and positive feedback, to be used to feed-forward to later progress. For it to foster intrinsic motivation it must:

- as already noted, reflect generally successful learning; and
- provide detailed diagnosis of success and failure, as information rather than as reward and punishment

Again, the reinforcing of success concept does not mean 'going soft'. There needs to be correction of errors, clarification of meanings and removal of misconceptions individual by individual. Such diagnosis was, as Taba and Elkins stressed in a book on *Teaching Strategies for the Culturally Disadvantaged* (1966), 'a continuous and perennial task' (p.66).

Feedback in the form of detailed comments is required, therefore, discussed personally by teacher and pupil. More time will need to be spent with weaker pupils. How is this to be found? Two possible ways are to expend less energy on formal class teaching and in mass marking of scripts. Flexible learning (*see* above) may help in these ways. If marks have to be assigned, it is preferable that they should be related to the individual's own potential, and not aggregated at the end of term to allow class rankings. Apart from its degenerative influence on the pupils lower down in the form, it may well be that a pupil of a particular ability and attitude near the top of a weak class may be underachieving, while one near the bottom of a strong class may be working her or his heart out. This crucial pedagogic principle is one to which the current government appears particularly insensitive, in their attachment to publicising league tables. It is keener on using the information less for constructive feedback, and rather as an accountability measure of teachers and whole schools.

Summary of Guidelines

The practical implications of the various insights into pupil learning and teacher influence considered in this chapter can be summarised in the following guidelines, applied to geography, presented as likely to promote good educational practice. They are not, of course, conceived of as catch-all solutions:

1. In the early stages of the secondary school course, the content should be rooted as far as possible in reality, starting off with matters within the experience of the children.

2. As other criteria dictate that not all work should be confined to the local environment, a concentric structuring of the syllabus is valuable, exploring vital principles, as the National Curriculum permits, first in the home area, before extending these outwards to the national and global levels. The previous use of contrasted locality study material in the primary schools should prove helpful in supporting this process of decentration. It must be stressed that concentric here implies the process taking place theme by theme, so that decentration comes quickly.

3. To infuse 'familiarity', place studies should as far as possible be presented as 'fieldwork in the classroom', with the help of visual and other first-hand materials.

4. Perceptual and conceptual problems have to be overcome in atlas, map and photograph interpretation. Here, too, introduction of maps and photographs of familiar scenes should precede those of the unfamiliar. Large-scale maps should be introduced before small-scale ones.

5. For a number of reasons, including compensatory ones so far as girls are concerned, the traditional over-pervasion of map reading skills in Year 7 work should be ameliorated.

6. As far as possible, independent learning should be catered for, for example through the use of worksheets, though it is important that worksheets should be seen as only one of a variety of appropriate devices. Worksheets can well reflect the core material of work units, in addition to which enrichment activities need to be provided for more able pupils, and reinforcement activities for less able. They must

be of a high quality in terms of content and presentation.

7. The nature and purpose of the work should be clear to the pupil. At all times it should carry meaning through stimulating concept formation.

8. Due attention should be paid to language issues, covering both vernacular and technical concepts and the functional links between them.

9. Overloaded syllabuses should be avoided, as adequate time is necessary to make possible meaningful learning. More limited, in-depth approaches are to be preferred, at the same time ensuring a balanced geographical input as between skills, places and themes.

10. At the adolescent stage, considerable effort should be devoted to fostering critical, explanatory thinking, through activities which can diagnose whether understanding, application and problem-solving tasks have been successfully accomplished.

11. Concepts should be revisited at progressively higher levels as part of a spiral curriculum.

12. Spaced assessments related to short units and blocks of time are advantageous in helping to fix concepts that need to be retained over a considerable period.

13. Individual differences must be recognised and catered for, in the full range from educationally backward to gifted pupils.

14. Positive motivation of the pupil through both the use of interesting starter-material, but more particularly through the promotion of successful learning, is critical.

15. Diagnostic feedback to pupil and parent, giving precise information about the quality of performance, is also a vital factor in motivation (unless it serves purely to reinforce failure), and should be achieved through regular and constructively critical reporting on progress.

16. Teachers who exhibit warmth and orderliness, and provide a democratic but structured and orderly classroom atmosphere, are more likely to promote favourable attitudes than authoritarian or over-permissive teachers.

17. Flexible teaching approaches, if implemented in cautiously critical way, hold high potential for improving student learning, and are not necessarily incompatible with National Curriculum demands.

CHAPTER 7

Assessment, Testing and Examinations

The National Curriculum: a Throwback to the Past?

Assessment in education 'is the process of gathering, interpreting, recording and using information about pupils' responses to an educational task' (Harlen *et al.*, 1994, p.219). There would be general agreement that the National Curriculum is assessment-led. The Secretary of State at the time of the 1988 Act, Kenneth Baker, made no bones about the fact that attainment targets would represent clearly specified objectives for what pupils should know, understand and be able to do when they reached the ages of 7, 11, 14 and 16. It was essential that attainment targets should provide a sound basis for assessment and testing, and in turn 'for publicising and evaluating the work of the education service and its various parts in the light of the pupils' achievements' (*TGAT Report,* (Task Group on Assessment Testing) 1987, Appendix B, p.2).

It has been this aspect of the Education Reform Act that has provoked the greatest controversy. Progressive voices have claimed that the National Curriculum is a throwback to the Revised Code of the 1860s. There are indeed alluring similarities between the two, but they are not nearly as straightforward as has been claimed. A brief historical perspective is demanded (Daugherty, 1990). One major similarity was in the attitude to what different generations have termed the 'basics of education'. During the 1850s, government expenditure on provision of elementary education was soaring and the impression was gained that its scope had become too ambitious, resulting in a neglect of the so-called three R's. There was the suspicion of a conspiracy of teachers, and even HMI, against the interests of the majority of the children and the state exchequer. Inspectors and teachers were accused of concentrating on the state of the upper rather than the lower standards in schools: 'The...danger is that the grant for education has become, instead of a grant for education, a grant to maintain the so-called vested interests of those engaged in education...' (Hansard, 1861, Col.211).

The Revised Code of 1862 harshly tightened the system, making government grant depend on success in the annual examination and on good attendance. It was indeed motivated by attitudes similar to those associated with the implementation of the National Curriculum, in that it was deliberately intended as a means of assessing the achievement not only of pupils, but also of teachers and schools, inflicting great damage on teacher morale. At the same time, teachers quickly learned how to play the new games, which largely involved effective parroting of information by pupils on the day of inspection. As Matthew Arnold put it, 'in the game of mechanical contrivances the teachers will in the end beat us...'(1867, p.115).

At about the time this new assessment system was being imposed on the elementary sector, so the secondary schools, increasingly demanded by a growing middle-class population, saw how important the institution of external appraisal was to their credibility. Public examinations were regarded as one of the great inventions of the nineteenth century, in that they substituted a meritocratic for a nepotistic system of occupational and social advance. Merit meant ability plus effort, and the hard-working children of respectable homes were seen as deserving of upward mobility.

A second major advantage of these examinations was thought to be the incentive they gave to pupils to work hard, success materially affecting 'their future advancement and prospects of life' (Booth, 1857, p.22). The examinations were successively introduced to raise the standards of entrants to the Civil Service and, later, to the universities and other walks of life.

Much current government thinking can logically be attached to all these Victorian innovations, though it can be argued that the Thatcherist spirit is more to do with the return to the values of a meritocracy than the primitive crudities of payment by results. Having said this, there is a disturbing similarity in the attitudes towards the educational professions that was articulated by Robert Lowe and has been revived by some of his successors of nearly 150 years later, not least in the 'league table mentality'.

The Task Group on Assessment and Testing

The official Task Group on Assessment and Testing (TGAT) did not, however, subscribe to the extreme demands of the league table adherents. The Group's basic criteria were not inconsistent with good practice. It was made clear, for example (1987, para 4), that assessment should be the servant and not the master of the curriculum. The criteria of good assessment practice were presented as follows:

- The assessment results should give direct information about students' achievement in relation to objectives: they should be criterion-referenced.

- The results should provide a basis for decisions about students' further learning needs: they should be formative.
- The scales or grades should be capable of comparison across classes and schools, if teachers, students and parents are to share a common language and common standards: to this end the assessments should be moderated.
- The ways in which criteria and scales are set up and used should relate to expected routes of educational development, giving some continuity to a student's assessment at different stages: the assessments should relate to progression. (para 5)

Over-arching the criteria was the conviction that assessment had a central role in education:

'Promoting children's learning is a principal aim of schools. Assessment is at the heart of this process. It can provide a framework in which educational objectives may be set and pupil's progress charted and expressed. It can yield a basis for planning the next educational steps in response to children's needs. By facilitating dialogue between teachers, it can enhance professional skills and help the school as a whole to strengthen learning across the curriculum and throughout its age range.' (para 3)

Purposes of assessment

- *Formative* – So that the positive achievements of the pupil may be recognised and discussed and the appropriate next steps may be planned
- *Diagnostic* – Through which learning difficulties may be scrutinised and classified so that appropriate remedial help and guidance may be provided
- *Summative* – For the recording of the overall achievement of a pupil in a systematic way
- *Evaluative* – By means of which some aspects of the work of a school, an LEA or other discrete part of the educational service can be assessed and/or reported upon.

Clearly for teachers, in a context of over a decade of distrust of government policy, the sting was in the tail. Indeed, notwithstanding the educational probity of most of the TGAT recommendations, they caused much trouble to come. One reason for this was their complexity and the consequent problems the subject groups had in translating the TGAT formulae into good curriculum and assessment practice. More fundamental was the feeling of most practitioners as well as theorists in the world of education that the intentions and interpretations of the TGAT recommendations by government were both reactionary and punitive in intent. The government as a policy priority introduced a high-stakes element into the proceedings, which meant that teachers' commitments to their pupils would be compromised by an understandable need to narrow their horizons and play for their own safety. This confusion of purpose was not avoided by TGAT in juxtaposing largely contradictory principles, like using the same test instruments for formative assessment purposes and for teacher and school appraisal. What good assessment practice demands is a clarity of purpose, with different assessments used to meet different objectives.

The intentions and also the fundamental grasp of the principles of good practice by TGAT were also subverted by the indecent haste with which it had to produce its report. In consequence, there were hasty judgements made, misunderstandings, conflicts, and contradictions. The complications of the recommendations led members of the School Examinations and Assessment Council (SEAC) from an early stage publicly to claim that the TGAT formulae as being interpreted by the subject groups could not be implemented in practice. The later dramatic volte-face, in the face of the bureaucratic nightmares which resulted from the numerous Statements of Attainment and associated SAT arrangements, demonstrated these claims to be well founded. The government, in setting up the NCC and SEAC as separate entities, additionally contradicted a key TGAT principle of good practice: namely that curriculum and assessment are inseparable. TGAT was finally in an impossible position as it became crystal clear that its recommendations were to be harnessed and if necessarily adulterated in the interests of pre-conceived government policy.

I shall now look further into criteria of good practice in assessment, as recognised long before the National Curriculum came on the scene.

Types of Assessment

As TGAT rightly asserted, a prime distinction is that between what have been termed summative and formative assessment procedures.

Summative assessment is most obviously represented by the external examination, but would also include internal end-of-term or end-of-year tests which function as pilots for an external examination. The SATs of the National Curriculum are examples of summative assessment. This form of assessment is likely to be **norm-referenced**, meaning that it is related to a set of norms, or average performances, drawn from an appropriate population, which act as a kind of bench-mark against which the pupils can be measured.

Formative assessment is a monitoring procedure undertaken on a continuing basis. It is **criterion-referenced**, meaning that the pupils are expected to attain appropriate levels of knowledge and understanding. The aims provide the criteria, and all should be able to 'pass the test' if sufficient effort is made. Its less formal and continuing nature may even persuade teachers that they are not assessing at all. While to many in education formative assessment is more congenial than summative, it has also to be accepted that there can be good summative assessment practice, just as there can be bad formative assessment practice.

The Purposes of Summative Assessment

Placing or Grading of Candidates

The first intention is to measure the candidates' achievements as accurately as possible, so as to place them in a rank order. The measure may be expressed as marks or grades. The placing may serve some specific vocational goal, such as selecting a limited number of candidates in a competitive situation, in which case the number of 'passes' is constrained by the number of places available. This competitive type of test is different from the qualifying GCSE type, in which large numbers of candidates can pass if they achieve the standards laid down. A pass result is taken as evidence that the candidate has satisfactorily completed a course of study. The selection of the pass/fail line is made on a normative basis.

Predicting Future Achievement

This second type of placing also has a predictive function, for it is held to qualify a candidate on the upper side of a pass/fail line, in some relatively diffuse way, as fitted to undertake a wide range of future activities. A high placing is presumed to be diagnostic of greater chances of later success than a low placing. The examination, therefore, acts as a clue to future achievement. Prediction is likely to be more accurate where future activity is linked with previous learning. A high 'A'-level grade in geography, for example, is thought to predict a good honours degree in that subject. But there is not a direct correspondence, as other variables are involved, such as the personality of the student and the methods of teaching adopted in the school. Pupils have been known to over-achieve in school, where they might have been too carefully nurtured, and under-achieve in the more *laissez-faire* atmosphere of a higher education institution.

Equally problematic is the situation where a 'transfer of training' is needed. Possession of a good honours degree is no guarantee that a graduate in geography will make a satisfactory teacher of the subject, nor a good town planner, nor a successful manager in industry. On the other hand, a string of good 'A' levels is usually at least diagnostic of hard work. If an employer is looking for 'clerkly diligence', s(he) might find conventional examination success reasonably predictive of the qualities required. Those whose demand is for originality and initiative might well find that an old-time examination process has militated against the development of such attributes.

The Purposes of Formative Assessment

The major purpose of formative assessment is the monitoring of pupil progress. It is designed to provide continuing feedback on the progress of the individual pupil and of the class as a whole, diagnosing weaknesses and suggesting appropriate remedies. While summative assessment provides a once-off, global feedback, that from formative assessment is regular, detailed and precise, so long as the appropriate instruments are used. Its twin advantages are therefore feedback to the pupil and adaptive feed-forward to teacher, for future planning on the evidence of the formative assessment.

Two other functions are served by both formative and summative assessment: to provide incentives and raise standards; and to assess the effectiveness of teaching.

Providing Incentives and Raising Standards

Assessment procedures remain in many circumstances a stimulus to effort on the part of both pupils and teachers, and thereby a means of raising standards. Summative assessment encourages extrinsic motivation, in that a long-term reward in the shape of improved occupational prospects beckons. While we would prefer to think that motivation should be intrinsic, it would be flying in the face of generations of evidence not to accept that summative assessment has a motivational impact on many pupils. It may well be that formative assessment is more likely to foster intrinsic motivation than summative, but this is only likely to be the case if regular formative assessment promotes the success that gives confidence and kindles interest. Achievement motivation should not be downgraded as a negative force leading to an unacceptable success. Successful achievement is, as we have seen, crucial to the cognitive growth and self-esteem of individual pupils.

Assessing the Effectiveness of Teaching

The teacher must expect, within reason, to be held responsible for the progress of pupils. In the case of formative assessment, the feedback on success or otherwise will in the first place be personal to the teacher, though peers and parents will undoubtedly be making perhaps very subjective appraisals on the basis of the informal evidence. In the case of summative assessment the evidence is public, and likely to be regarded as testimony by both headteachers and parents as to the success of the teaching. As in the payment by results period, so today, insensitive handling of this delicate issue is calculated to provoke discontent. The major problem is that teachers are not all working on a level playing field, and some experience much greater difficulties than others in handling and motivating children, depending on the school situation. If league tables of schools are to be produced, it is logically and morally essential that 'value-added' techniques are used to adjust raw scores, because most of what the current scores are reinforcing is the long-standing truism that academic success is linked with social background. Bad assessment practice is thus being promoted, offering distorted if not false evidence.

Good Practice in Assessment

The Pupil

To take the pupil first, the system must not serve to reinforce failure, and particularly early failure. In the internal context, while there is every need to monitor progress and be critical of individual deficiencies, it is worth repeating that it is damaging to publicise overall class rankings. Apart from the fact that they reinforce a sense of failure for many pupils, as already noted, a child near the bottom of a good class might be working nearer her or his optimum than one near the top of a weaker group. It is generally an advantage to involve pupils in the assessment procedures, being at one with their teachers on the evidence the assessment affords, and negotiating pathways to future improvement. The satisfactory resolution of the intrinsic tension between making the assessment too easy, and thus easily encouraging pupils, and stretching their abilities through more searching questioning, is also a mark of good educational practice.

Another important aspect of formative assessment is to measure pupils not against external normative criteria, but in relation to what they have achieved before, which will indicate progress or regress from an earlier position. This has been termed **ipsative assessment**.

An increasingly popular way of involving pupils in the assessment process is through the practice of encouraging self-assessment, mentioned earlier in the context of individualising learning. This can be achieved through check-lists, evaluation sheets and in tutorial time. It is, however, difficult to standardise and meet criteria of reliability and validity (*see* below).

The Curriculum

A second major issue is that assessment should serve and not pre-empt curriculum planning. The assessment must therefore appraise a wide range of curriculum aims and objectives, and not be restricted to the mechanical testing of low-level knowledge and skills. Criticism of external examinations in the past has focused on the argument that they have represented the assessment tail wagging the curriculum dog. A key test of an examination which reflects good assessment practice is that it will not reward pupils who merely learn and regurgitate notes.

A Range of Techniques

Properly implemented, external as well as internal testing and examinations should be reconcilable with the broad aims of education, and play something more than a narrowly instrumental role. To do this the assessment instruments must be specifically related to these aims. Thus, if the aim is to develop critical and imaginative thinking, there is little point in relying on a monolithic assessment device that manifestly tests neither. An important part of the process lies therefore in extending the range of techniques of assessment. Assessment by means of once-and-for-all written papers may not only set a premium on literary ability, which it may not be the prime intention of the examination to test, but also favour a particular type of learning style. Thus summative assessments should not be based purely on **essays** or on extended pieces of writing such as reports; nor should they be restricted to objective, including **multiple choice**, items. The advantages of both can be combined in **structured or partly-structured questions**, taking in a range of techniques. For geography it is important to use in particular **data-response/stimulus questions**, in which data are offered to stimulate responses, so that students do not have to rely purely on the memory of what they have learned.

Criteria of Good Practice

However preferable formative assessment is, in most national systems it is impossible to escape the public demand for a summative element. Since the mid-nineteenth century there has always been this provision in England and Wales, and the practical issue is how to make it as supportive of good educational practice as possible. By definition, summative assessment needs to have external credibility. There are two key criteria in this respect: reliability and validity.

Reliability

The reliability of a test refers to whether it will give consistent results if repeated under similar conditions at a later date. It is a particularly important quality in public assessment, and is bound up with the discriminating power of the test.

- *Problems of scoring* – A key problem here for essays (and perhaps for course work) is that marking is a notoriously subjective and unstable procedure. Not only do different examiners mark to different standards: the same examiner confronted with the same piece of work at different times is also likely to be inconsistent. The assessment may be made more reliable if a second-marker system is built in, and a marking scheme agreed.

- *Attenuation of the mark range* – There is a general tendency in extended writing assessments not to use the whole range of marks available. If an assignment is marked out of 10, the subjective marker usually works within a spread such as 3/10 to 8/10, giving an effective mark range of 5. Such attenuation may not be important in formative assessment, but in a terminal test it lowers powers of discrimination and thus of reliability.
- *The effects of a limited number of questions* – Almost by definition, an essay test alone involves the candidate in answering a limited number of questions. The effect of this limitation is again to lower powers of discrimination. The use of a range of other techniques will increase the range of questions and raise the reliability. The exclusive use of assessments involving a very limited number of questions must make differentiation difficult. While there can be differentiation by outcome in such assessments, it is more difficult to build in differentiation by task, or by stepped questions or inclines of difficulty. These require the scope of a range of assessment modes and a large number of questions.

Validity

In educational terms, validity is an even more important quality in a test than reliability. Does the test achieve what it sets out to achieve? Two main types of validity have been identified:

- *Content validity* refers to whether the content of the test is adequately related to the teaching and learning that has previously taken place. It refers not only to factual content, but also to the more general objectives of the course.
- *Construct validity* refers to whether the test fairly measures the attributes of the candidate which it is ostensibly designed to measure.

Content Validity

Content validity is influenced by a number of factors:

A Limited Number of Questions

Apart from diminishing a test's discriminating power, a limitation in the number of questions means that it cannot adequately sample the syllabus. It may well be that a large number of questions will be set on the paper to cover the syllabus reasonably well. But the candidate can only choose four or five of these, which do not necessarily reflect the work s(he) has undertaken.

Choice of Questions

The fact that the candidates are given this choice means that they are effectively sitting different examinations. The degree to which prior coverage will match the questions set will vary from school to school according, for example, to whether the teacher determines to cover the syllabus at all costs, however superficially, or whether s(he) chooses the educationally more desirable in-depth approach.

What Ability is Being Tested?

Perhaps the most crucial aspect of content validity is whether the test accurately reflects the objectives of the course. Let us assume these require the test to do more than assess the ability to recall facts. Extended writing modes are sometimes still presented as the most effective means of assessing higher level abilities, such as synthesis and imaginative response. This point is accepted, providing that the questions genuinely do demand these qualities and, more important, that this is recognised in the marking scheme. Often pieces of extended writing involve merely the recall of teacher notes.

Construct validity

Various factors conspire to make it difficult also for an extended writing type of examination to achieve construct validity.

Difficulty Levels of Questions

As has been noted, any significant choice element means that different candidates are sitting different examinations, even though the same papers. This is not only the case in respect of prior content coverage, but also because of the varying degrees of difficulty of questions. Some candidates will make a wiser choice than others, thus introducing a further variable into the situation. The examination becomes a lottery, destroying its validity.

The Wording of Questions

Many extended writing type questions are vaguely or ambiguously worded. In a situation where so many of the command words such as 'analyse', 'assess the importance of', 'discuss' or 'describe and explain' are translated down to an all-embracing 'write about' there may be safety in numbers. Where words do mean something, however, and those in the question have some influence on the marks being allotted, differential understanding by candidates of ambiguities in questions may well mean marks being awarded for accomplishing different tasks. Where this occurs the marks are strictly non-comparable.

Communication Skills

The basic purpose of a geography examination is to assess achievement of geographical understanding. Yet it is difficult for an examiner in marking essays to ignore such 'non-geographical' qualities as contrasts between the tidy and untidy, the grammatical and ungrammatical, compact and rambling work and so on. The situation is even more delicate in internal examining, where the pleasant and co-operative pupil starts off with a halo round her or his script. It is not being argued that cosmetic qualities should not be assessed, but that they should be assessed separately, and certainly not several times over in the same examination.

This currently is an issue between teachers and Secretaries of State, who appear to favour what is bad practice, that is making every examination into a spelling and/or punctuation test. This is not to say that these are unimportant skills, rather that a dedicated test is needed to assess them.

Culture and Gender Bias

Essays have a long culture, a distinct gender, and also a social class bias. Over-emphasis on the essay has had the pernicious effect of loading the dice heavily against certain groups, exacerbating the difficulties of so-called 'socially disadvantaged' pupils, and also of children who were slow readers and poor writers. Prose-writing is not a form of communication which comes naturally to them and the disadvantage has been hammered home by repeated use of the procedure in different subjects in the same examination. It is difficult in these circumstances to decide whether deficiencies have arisen from lack of innate ability in the subject, or from problems of communication in 'standard English'. In this way again the essay is not fairly measuring the basic attribute it was introduced to measure.

On the other hand it has been argued that essay-type assessments may be of advantage to girls who are said to work more diligently and neatly, and be more responsive to this traditional form than, say, to multiple choice tests, in which boys are claimed to achieve better. The gender issue will be referred to again later.

Assessment by Final Examination or by Course Work?

Many teachers would maintain that there are great advantages in concentrating on the teacher assessment of course work, rather than relying too much on the outcomes of external examinations.

Advantages of Assessing Course Work

- Course work assessment is a vital element in building up a cross-profile of the pupil. A well-balanced assessment equally requires the use of a range of techniques. In geography the course work will often take the form of a detailed field work project, or a similar venture based on first-hand library investigations.
- This type of assessment may shed a new light on the work of those who are not at their best in a situation where rigorous time constraints operate, as in the conventional examination. It caters for different learning styles, can provide due reward for the offerings of both convergent and divergent thinkers, and may also be used to assess a greater variety of communication skills than the purely verbal.
- Course work offers unrivalled opportunities for creative presentation, including the use of Information Technology.
- Assessment of course work is probably the best means of providing evidence of the candidate's creative abilities and skills of communication, and the degree of benefit or otherwise derived from enquiry-based learning. It is particularly helpful in geography also because it allows for the consideration of work related purely to the field studies, which is difficult in a conventional written paper.
- The knowledge that course work will be assessed is a stimulus to effort over a period of time. More important, there is considerable evidence that, even among pupils whose conventional work is undistinguished, the interest of a properly guided, in-depth personal field study often takes over, leading to enhanced pride in work and consequently higher standards than are manifested in other parts of the assessment. Field work is obviously not the only type of course work. There is, however, much less to be said for examining externally course work in the form of a file built up of notes made in lessons and other material based on second-hand sources. The assessment of such work should be formative, providing continuing feedback inside the school.
- Course work assessment is an essential element of flexible learning, which is by definition an ongoing process.

So long as the project is properly chosen, course work assessment has educational advantages that are difficult if not impossible to replace by other modes. Apart from advantages accruing to the pupils, such work is almost invariably more enjoyable to mark than the mass examining of essays, or the tedious if less onerous task of checking multiple choice items.

Problems of Assessing Course Work

- In a nationally or regionally operated system, course work is likely to be summatively assessed. A major problem lies in ensuring standardisation between schools. This is usually achieved by a process of moderation. After the preliminary internal assessment, moderation may be undertaken by staff drawn from local consortia of schools, a practice adopted by a number of examining boards. At a later stage, samples of course work are moderated by qualified subject specialists appointed by the board.
- In internal marking, it is difficult for the teacher to dissociate her or his view of the pupil from the pupil's work. There can easily be a hidden mark for co-operative attitudes and neatly presented work, as we noted with the essay. Clearly course work is likely to involve an element of extended writing.
- There is also a tendency to bunch marks in internal assessment, which is particularly the case in field work projects, where it is less likely that much work of an exceptionally low standard will be submitted. The project might be looked upon as an insurance policy for the weaker candidates. On the positive side, the interest this type of work generates will also tend to push up standards.
- There is additionally the danger of judging a project rather in terms of its weight and cosmetic appearance than its inherent geographical qualities. Thus it is important that the criteria for marking are carefully laid out, with separate marks allotted for presentation. Even if no direct penalties for irrelevance are imposed, some sort of procedure is needed to ensure that work that is compact and to the point is ranked higher than diligent but rambling accounts. But relevant points made must be credited, even if hidden in the bracken of irrelevance.
- A related problem is posed by differences in quality of teaching and teacher expectation as between schools, and also within them, which can emerge starkly in the moderation of course work. The tensions are obviously not confined to this type of assessment alone, but can cause embarrassment and even acrimony at meetings where studies of highly disparate quality are being publicly appraised by local consortia of teachers; or, within schools, by members of a faculty.
- It may also be difficult to judge how individual the work being presented is. How much is there in it that reflects the efforts of the rest of the peer group, of relations and friends and, especially, of the teachers themselves? Guidelines are laid down by some examining boards, but all must rely in the last resort on professional integrity and on the honesty of pupils and their families. The main problem in geography field work projects is that some of the objectives and methods can best be achieved by group study. In geomorphological field work, for example, the measuring of slopes demands the co-

operation of more than one person. Teachers are likely to be anxious about the reliability and validity of their assessments unless clear directives can be given on this issue and threshold cases are decided. Obviously, the course work requirements are a big advantage to pupils from supportive and well-to-do families, and this in itself rouses suspicions about the fairness of them. They have also been said to favour girls, perhaps redressing the balance with respect to other forms of assessment (Gipps and Murphy, 1994, pp.218–21).

- Teachers can also find it difficult to cope with the internal administrative problem of organising and assessing large batches of assignments for whole classes of pupils. There are also financial constraints. There is the added tension of the potential 'drowning' of local farms, villages, shopping centres, and fragile environments, by hordes of juvenile researchers. Then there is the burden of moderation. The working of consortia, though a valuable means of co-operation between teachers, is none the less another time-consuming exercise.

Approaches to Course-work Projects

- The topic should be on a locality scale, whether pertaining to the local area of the school or a more distant place. If parents agree, there is no reason why the choice could not be of an area visited on holiday, or where relatives or family friends live. The small-scale, in-depth study, such as an analysis of a bend in a local stream, usually leads to more interesting and personal work than grandiose projects such as 'Aspects of the Geomorphology of the Wye Basin,' where the content tends to be second-hand and cumulative.
- Although the topic should clearly be geographical, it would be a pity if this were too narrowly interpreted. While the norm would be a field work project it would seem quite legitimate to study, for example, the changing patterns of journeys to work, based on research in the library using census materials, rather than field evidence. There are also many interesting historical geography topics that might necessitate concentration on library research. The 'geographical test' is not whether studies are undertaken in the field, but whether the processing and 'placing' of the information is distinctively geographical. The new Orders fortunately do specify secondary as well as primary sources.
- The material should, of course, be collected at first-hand as much as possible. Thus reliance on brochures from industrialists, travel agencies and the like, though perhaps useful in providing illustrative detail, should not represent the central resource base. Above all, the scissors and paste method of using such material, involving the copying out of verbal sections and pasting in of visual, together with a lack of annotation or 'translation' of the data contained in maps, diagrams and photographs, should be discouraged. Where second-hand material is

used, it should be attributed. There is no reason why pupils from Key Stage 3 upwards cannot be taught, as part of the techniques of presentation, the skill of attributing material to its source.

- A wide range of methods of presentation should be encouraged, maintaining a balance between verbal, numerical, graphical and cartographical forms, maximising the use of information technology. Properly employed, this will do more than give an attractive presentation: it will help to spotlight the structure and sequence of the material.
- The issue of structure is very important. One recommended method is to use an enquiry, problem-solving approach. The process is as follows:
  - Start with a problem issue (one, of course, that is worth investigating). This might, for example, be to do with reasons for variation in the density and nature of traffic on the local by-pass, the main road through the village, and the minor road past the school.
  - Advance a hypothesis, such as one related to differences between internally and externally generated traffic, e.g. traffic is heavier on the by-pass than on the other two roads because it is mainly generated by two nearby larger towns which the by-pass serves to connect.
  - Decide what data needs to be collected in the field, e.g. traffic counts, a census of vehicle types, their place of origin if possible
  - Collect and roughly code this information, using base maps, observation, field notebooks, and so on, in a neat enough form for it easily to be interpreted later.
  - Classify, if necessary, and present the information on maps, graphs, matrices and so on. There is obvious potential for introducing IT here.
  - Describe, analyse and interpret the evidence which emerges, i.e. testing the hypothesis. (At more advanced levels, correlation will not be intuitive, and may be based on statistical tests of significance and so on.)
  - Consider whether the results justify accepting the hypothesis or not.
  - Show awareness of the limitations of the methods used, evaluating how appropriate they were to the exercise.

Clearly at earlier stages the teacher will tend to guide the children towards hypotheses that are likely to be supported by the evidence. The degree of rigour insisted upon for the acceptance of the hypothesis will also vary with stage of development. If the hypothesis falls, a new one might be set up and tested. If it is established, it might lead on to some further generalisation, about traffic patterns in this case.

This method is no panacea. It does not guarantee good work, and can become as stereotyped and mechanistic as any other method. Much depends on the generative power of the problem posed, and the quality of interaction between teacher and students. If the problem is not worthwhile solving, the product is likely to be no better than traditional accounts which merely accumulated verbal information. Creativity should not be stifled by a rigid insistence on particular methods of processing. There is no inherent reason why this scientific approach should inhibit creative work, but it must be stressed that there are other routes to enquiry than the scientific model illustrated here.

Oral Assessment

Oral assessment can be achieved through direct questioning, interviews, discussions, including discussions or debates within a peer group, or through more formal approaches in which students make presentations to a peer group, or through role play. The presentations need not be in the form of debates or drama, but can alternatively be through using the technology of audio-tapes and video recordings.

Oracy is clearly a vital cross-curricular skill. While arguably it may be more central to other areas of the curriculum than geography, enquiry-based, problem-solving activities in geography, in which discussion of the pros and cons of issues can be diagnostic of student understandings, have encouraged geography teachers to extend their interests in this area. In investigative activities, for example, pupils can be encouraged to:

- use talk to relate new information to existing experience;
- investigate, speculate, negotiate, speculate;
- argue, reason, justify, consider, compare, evaluate;
- confirm, reassure, clarify, select, modify, plan;
- demonstrate, narrate, describe;
- evaluate new understandings (Carter, 1991b, p.2).

Discussion of issues can, understandably, readily stray into exchange of prejudiced opinions, and it is the task of the teacher to establish such necessary conditions as focus, relevance and balance, before any attempts at appraisal are made. It is important for the teacher, as part of departmental or school procedures, possibly based on agreement trials, to have on paper clear indicators of achievement in oral work, and how in group work the individual contribution can be appraised. Otherwise the nature of the assessment may be highly subjective and idiosyncratic.

Some pupils who find writing difficult may be helped by the building in of oral assessments. To others, however, they may be threatening. Oral assessment is also time-consuming. It can again be urged that oral assessment is useful to include, but as one of a battery of different types of assessment.

The Assessment of Attitudes

As will be argued in Chapter 9, the values and attitudes dimension is central to the effective teaching of geography, bearing in mind the various principles advanced in earlier parts of this book. It is well known, however, that it is difficult to assess attitudes formally. Some might argue that it should not even be attempted.

Information on attitudes to the subject in the first place is needed. It is vital in at least three ways:

- in the direct sense that interest in a subject is essential to meaningful learning;
- because a teacher ought to have feedback on whether such aims as stimulating interest, awareness, appreciation, involvement and so on, have been achieved in some degree;
- because positive attitudes towards something may have more predictive validity than success in a written paper.

A certain scepticism regarding the assessment of attitudes is understandable, for they are relatively intangible and difficult to evaluate. An attempt to assess attitudes quantitatively on the basis of criteria such as willingness to co-operate in the normal routine of class work, persistence in pursuing the task, resourcefulness and enthusiasm should all be observed and recorded. But can giving extrinsic recognition through marks for these intrinsic qualities be justified? Further, favourable attitudes may be ephemeral. 'Liking geography' may really mean liking a particular geography teacher. The promotion of favourable attitudes is a basic aim of schooling, however. So it is important that they are appraised qualitatively and recorded, if not quantitatively assessed.

The attitudinal dimension is even more important in relation to developing positive attitudes towards the world outside. This aspect will be considered in Chapters 8 and 9. The relationship with assessment is again problematic. It is argued here that credit should not be given for uncritically following a particular party line, that may to an extent have been inculcated, whether on the side of the angels or not. There is a built-in dilemma. What happens if, for example, unacceptable attitudes in human terms are skilfully communicated and logically justified? How do we assess a technically brilliant piece of writing extolling racism, for example?

Profiling, Records of Achievement and Reporting

In 1986 Graves and Naish produced a useful publication for the Geographical Association on profiling in secondary geography, much of which still holds good. By the time it was written it had long been accepted that old-time reports to parents were minimalist and unacceptable, and that some more considered profiling was necessary as a record of achievement to be reported back to parents (*see* Hall, 1989).

In a profile, pupils, teachers and parents will have a record of a whole range of coverage and experience in different parts of the scheme of work, the quality of achievement, and elements of the attitudinal dimension, such as effort. Thus profiles of a more basic sort might variously start under traditional headings such as:

Effort	*Attainment*
Outstanding	Excellent
Good	Good
Satisfactory	Sound
Poor	Below average
Serious lack of effort	Weak

This in itself would be unsatisfactory in not going much beyond what used to be found in old-time reports. It would function as a mere starting summary. More vital would be the breakdown according to more detailed criteria, including, for example, criteria of:

- attitude
- initiative
- written communication
- non-verbal communication
- reference skills
- knowledge acquisition
- understanding
- problem-solving
- analysis
- synthesis
- evaluation
- decision making
- empathy

If a detailed breakdown is wanted, these criteria could be collapsed into sub-criteria, then each be graded on the kind of five-point scales above, before overall aggregation. Obviously to cover these different criteria and also link them with the different elements of enquiry-routed and criterion-referenced schemes of work in geography, can lead to the kind of bureau-cratic over-kill that the Statements of Attainment of the National Curriculum in geography were criticised for. Thus the different specifics of geography can also be divided; as with map skills:

- can use grid references
- can use a compass
- can interpret and use a key
- can interpret and use scale
- can understand and interpret contour maps, and so on.

An important quality of a good profile is therefore to avoid what the

Americans have called numerosity, while at the same giving clear and reasonably detailed information. Too much ticking of competency boxes is unfortunately increasingly being interpreted as progressive practice. We might look back at checklists of objectives to be secure that we remain in the arena of educational improvement, assuring ourselves that our objectives are worthwhile and are couched at an appropriate level of generality, as well as being clear and legible to pupils and parents as well as teachers (*see* Chapter 5).

The Northern Partnership for Records of Achievement offered useful advice on these issues (quoted in Lambert, 1990, p.36). It drew on the validation of a large bank of ideas and strategies from different schools and authorities, which helped to break away from the constant reinvention of the wheel which tends to occur in situations where pioneering schools and LEAs publicise their efforts, which might be useful, but at the same time appear to imply that their system is the one most generally applicable. No such best solution will be offered here.

An essential element of good practice in profiling is to offer regular reporting so that pupils are apprised of their progress or otherwise at frequent intervals. Another is to build in an element of self-evaluation. Many schools include this feature today. It characteristically will include reference to pupil's perceptions of:

- Coverage of content (places and themes);
- Range of skills acquired;
- Understandings of content achieved;
- Problems met with;
- Enjoyment of the unit;
- Evaluation of whether targets have been achieved;
- Agreement with teacher on future targets;
- Ways of achieving improvement in meeting these targets.

In the post-Dearing National Curriculum arrangements, hopefully teachers will now be more free to devise their own systems within their frameworks of good practice, as built up through their experience of the profiling movement (*see* Daugherty and Lambert, 1994).

Towards a Balanced Approach to Assessment

It is only too apparent that in the increasingly fractious debate over education in England and Wales over the last two decades, assessment has been a stalking-horse. Extreme positions range from those who declare that under almost any circumstances assessment is anti-educational, to those who would like a return to something like the equivalent of the Codes of the 1860s, the 11+ examination, and an unrelieved diet of summative pencil and paper tests and examinations.

There has been a discernible shift in assessment practice away from the latter towards the former position in the last two decades, and it is this trend which the government in its 1988 Act had clearly decided it wished to stifle. At the other extreme are those who with equally strident polemic chant their opposition to all aspects of formal assessment. Rather than offering two polarised possibilities, let us propose a three-fold categorisation, based on three ideologies, termed here official, centrist and progressive (Table 7.1).

Official	*Centrist*	*Progressive*
Norm-referenced	Choice of	Criterion-referenced
Summative	appropriate	Formative
Externally administered	range of	Internal
Closed-questions	instruments/ procedures	Open-questions
Teacher-dominated	for given	Learner-negotiated
etc., etc.	circumstances	etc., etc.

Table 7.1

The left and right columns are, in too much of the literature on assessment, presented as inherently irreconcilable. One extreme view states, baldly, that norm-referenced assessment is bad for the pupil and criterion-referenced is good; that summative assessment is bad, and that formative is good, and so on. As has already been demonstrated, however, summative assessment is in a different category from formative, and does not lie at the opposite end of a particular assessment spectrum. On the same spectrum are rather a range of from good to bad practices within each procedure, whether formative or summative. Good assessment practice in terms of validity, reliability, etc. can be related to each of these categories. This is not to assert that there are not genuine tensions between the types of assessment in the left and right columns above. Nor is it to suggest that we should not count as generally educationally preferable, for example, formative and criterion-referenced assessment, rather than summative and norm-referenced. But we also must recognise that there are reasons for having both, so that the key criterion of good practice is that we chose the forms of assessment appropriate to the particular task, as outlined in the central column and discussed in earlier parts of this chapter.

Section C

GEOGRAPHY AND SOCIAL EDUCATION

CHAPTER 8

Geographical Education and Social Stereotyping

What are Social Stereotypes?

One of the fundamental problems facing us as geography teachers is that we are dealing with the complexities of the world's environments and the world's peoples. Inevitably we have to simplify, in part by a high level of selection, and in part by generalising about complex issues. Good generalisation is indeed an advanced intellectual activity. But there is also a downside to these processes of simplification and generalisation.

The act of generalisation, while essential to making sense of the world, carries with it the problem that in every generalisation there lurks a stereotype. A stereotype was originally a term used in the printing industry, meaning casting in a fixed form or plate. Translated into the wider world, it has become a metaphor, suggesting perpetuating something in a rigid form. In the social context stereotyping is often manifested in rigid and negative attitudes towards other people(s). Negative stereotypes have the following characteristics:

- They usually present **false and/or distorted information**, even though that information may well contain an element of truth, and may appear plausible and disarming.
- They generally reflect **over-simplified and undifferentiated thinking**: clear-cut, basic, 'no U-turn' judgements, as distinct from thought processes which can accommodate refined shades of meaning and ambiguity.

- They are usually **group-related**, favourable towards the in-group, or those supporting it (the 'good guys') and unfavourable towards out-groups (the 'bad guys').
- They are often marked by extreme **attitudinal rigidity**, often deriving from hardened cultural attitudes which go back for generations. The prejudices embodied are not readily dislodged by counter-evidence or logical argument. They can become more firmly fixed where reinforced by techniques of inculcation and indoctrination, and demagogic appeal.
- Hard-core social stereotyping has been associated with the defence mechanisms of people who have been deprived of adequate emotional nourishment and may indeed reflect serious **personality problems** in the affected person (*see* Fishman, 1956, pp.31–5).

Visual stereotyping is often in the form of a caricature, as found in a cartoon.

Racial Stereotyping

While negative stereotyping of other peoples did not begin in the nineteenth century, the opening up of the world and technological improvements in communications greatly enhanced its impact. As education became universalised, more and more young people were made aware of different places and peoples, either in school textbooks or in the burgeoning children's literature. As they became adults a vast array of journals, magazines and newspapers brought the wider world into people's lives. Most of the presentations of alien peoples were unfavourable, though ranging markedly in their degree of disapproval. They were usually based on differences in appearance, race and religion. Racism has long been the basis of some of the ugliest forms of stereotyping, and has the following characteristics. It:

- fixes the group characteristics of human beings according to rigid lines of descent; that is, on a hereditary basis;
- makes physical peculiarities the key signals of human differences and imputes to them a social and moral significance;
- raises the emotional stakes by locating these characteristics in a hierarchical social and moral ranking;
- reinforces these further by warnings of the perils of crossing blood and of the racial impurity degeneration it infers will inevitably take place if such occurs;
- has historically used this threat to legitimate conquest, coercion, exploitation and, in extreme cases, genocide.

There is thus a common thread of determinism and even of predestination in racist stereotyping. It has been justified by appeals to religion, on the basis of statements in the Old Testament, and science. In the nineteenth century many of those regarded as barbaric were also, in the Christian sense, heathens. The opening up of Africa in particular led to a twin thrust of imperial and missionary annexation, the object to conquer and convert the heathen: 'the darker the picture of African barbarism, the more necessary the work of the missionary' (Rich, 1986, p.4).

The buccaneer imperialist was the most avid promoter of late eighteenth and early nineteenth century racism, at this time still engaged in ensuring the continuation of slavery. Black people were characterised as 'brutish, ignorant, idle, crafty, treacherous, bloody, thievish, mistrustful and superstitious' and as 'wisely fitted and adapted to certain uses' by the ordination of the 'Divine Fabricator' (Long, 1774, p.354 and p.375). This late eighteenth century stereotype of the primitive is not much ameliorated in Keane's Cambridge University Press text of 1920, defining the Sudanese negro as 'sensuous, indolent, improvident; fitful, passionate and cruel, though often affectionate and faithful; little sense of dignity and slight self-consciousness; hence easy acceptance of the yoke of slavery; musical' (p.16).

A more sinister trend in the aftermath of the debates over Darwin's *Origin of Species* was the emergence of a number of pseudo-sciences, which in principle and practice promoted racist views. Craniologists, phrenologists and physiognomists inferred that from physical attributes could be interpreted human character and potential. As Reade, one of the more extreme proponents of racism put it (1863): 'it has been proved by measurements, by microscopes, by analyses, that the typical negro is something between a child, a dotard and a beast' (p.509).

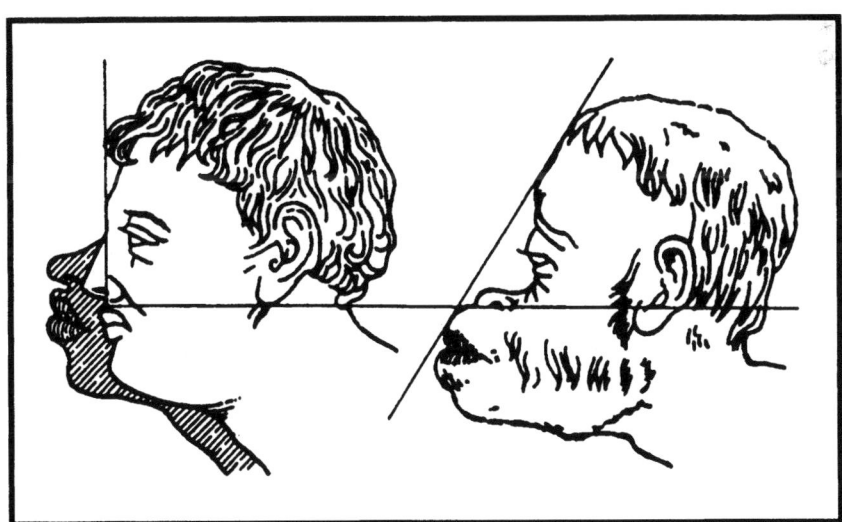

Figure 8.1 'Faces of Negro, European and Orang-o-tang'

Figure 8.1 illustrates a not untypical example of the application of these so-called scientific techniques, which were to extend Darwinist principles into Social Darwinism: the idea that there exists a preordained and hierarchical chain of being from beasts to Caucasian man, with negroid races placed nearer to the beasts than, for example, classical Greek types. One version of social Darwinist principles described five racial groups:

'1. **Caucasians**, such as Europeans and Arabs. Their fair blushing skin, high forehead, thin nose, long and soft hair, and great variety of expression, mark this most favoured kind. These white men have settled in every quarter of the globe, and have usually become rulers.

2. **Mongolians**, who dwell in China, Japan, and the Arctic regions. An olive or yellow skin, strong hair, slanting eyelids, and high cheek bones are their main features.

3. **American Natives**, whom a red skin, long skull, and aquiline nose easily distinguish. Their many tribes are quickly dying out; their greatest enemies being spirituous liquors, which they call "fire-water", and diseases brought in by Europeans.

4. **Ethiopians**, are at once known by a black skin, woolly hair, projecting jaws, a broad nose, long heels, and flat feet. These points do not give any reason why we should not love the negro as a brother, and among the next, are more lowly groups of mankind.

5. **Australia**, and the islands near it, are peopled by Malays and a kind of Negro, nearly as black as that from Africa. The latter have narrow foreheads, flat noses, wide mouths, low stature, and weak legs. It is harder to awaken the belief in God among them than among any other nation on the earth.' (Mapother, 1870, pp.9–11)

Implicit in the thinking was that for both scientific and God-given reasons, the white races were destined to rule over the coloured races, and that they must protect their purity by avoiding inter-breeding.

'Out of the prehistoric shadows the white races pressed to the front and proved in their myriad ways their fitness for the hegemony of mankind.... Two things are necessary for the continued existence of a race: it must remain itself; and it must breed its best.' (Stoddard, 1920, pp.299–300)

The implanting of racial prejudice in the young was successfully achieved by the writers of 'wholesome periodicals' such as *Boy's Own*, and in the vast literature of children's story books by popular authors such as Ballantyne, Henty, Marryat and Kingston. Kingston was second only to Dickens in popularity among boy readers. His underpinning philosophy was crystal clear:

'As far as it is permitted to man to comprehend the decrees of the Almighty, we have reason to believe that the Anglo-Saxon race has been awarded the office of peopling the yet uninhabited portions of the globe, of spreading the arts of civilisation, and more than all, of promulgating the true faith of Christ among the lands of the heathen.' (1849, p.6).

Racism and Imperialism in Geographical Education

Much of the racist stereotyping was faithfully reflected in the writings of geographical educationists. Just as the character of the individual could be inferred from the countenance, so could national characteristics be interpreted from the appearance and culture of the inhabitants. But more powerful than the negative stereotypes of foreign peoples were the favourable hero-worshipping images of the British. To Miss Sturgeon of the Manchester Geographical Society, it was vital, for example, to make 'missionary geography' part of the syllabus, to trace the 'daring journeys' of these 'noble pioneers of the gospel'. She was equally keen on following the news from the ubiquitous foreign fronts, noting that a geographical knowledge and intelligence allowed the initiated vicariously to 'go side by side with them (the British troops) from one battlefield to another' (1887, p.85).

The association of such names as David Livingstone with the Royal Geographical Society gave it enormous prestige, and a key position in the spread of geographical education. The missionary, the explorer and the traveller was the 'real geographer...indifferent to whether he plunges into the burning heats of tropical deserts...or endures the hardships of an Arctic climate...he feels delight in the reflection that he is upon ground hitherto untrodden by man' (Hamilton, 1838, pp.xxxix–xl).

Predictably, the RGS dissemination was not disinterested. Thus one of the great names in the history of the Society, mountaineer Sir Douglas Freshfield, in defining 'The Place of Geography in Education', was unblushing in his chauvinism:

'Shall we English who inherit so large a part of the world not acquaint ourselves with our inheritance...? What has been the fate of our race? To be the greatest rulers and merchants and colonists the world has seen....Do you think we are educating children for this high destiny...by comparative ignorance of the earth's structure, of the natural laws by obedience to which they may go forth and win peaceful victories and fill up the void spaces of our planet?' (1886, p.701)

Sir Halford Mackinder was at least aware there was some tension between disinterested science and political expediency, but was quite clear about the priority for education: 'Let our teaching be from the British standpoint, so that finally we see the world as a theatre for British activity' (1911, pp.79–80). Another educational writer, James Yoxall, in a late nineteenth-century geography text for pupil teachers, pulled no punches in perorations that the would-be pedagogues were expected to memorise:

THE EMPIRE. The United Kingdom is the ruling centre of vast possessions; valour and enterprise, conquest and colonisation, have won to British sway an Empire on which 'the sun never sets'. Our Empire extends to every zone and parallel, and shares in every continent and sea. It comprises one-sixth of the globe and one-fourth of the human race (1891, p.30).

Even the progressive Froebelian journal *Child Life* followed the imperial line. In advising teachers as to how they should deal with issues associated with the Boer War, it argued that children had the entitlement to learn of the cost of war and the national sacrifices which had to be made: '...the war has brought this people together as nothing else could have done; and the children surely have a right to share in the loyalty and patriotism which is stirring a great Empire' (1900, pp.50–1).

Nearer home, a negatively stereotyped group was the Irish. The issue was a sensitive one in Liverpool, so close to Ireland and with a large Catholic population. A member of the Liverpool School Board in 1900 complained that through certain textbooks being used in Liverpool board schools the children were being taught that the Irish people were 'ignorant' and 'wanting in energy', among half-a-dozen other degrading characteristics, all this being inculcated in schools supported by the Irish ratepayers of the city. He moved 'that the School Management Committee be instructed to substitute another book for Longman's Standard IV Geographical Reader, as this reader contains statements which give rise to disparaging remarks by teachers about the Irish people' (*School Board Chronicle*, 1900).

Even James Fairgrieve, a revered geographical educationist of the inter-war period, wrote textbooks and political geographies that were undeniably racist: 'Thus Africa, long occupied only by barbarous peoples...has lately naturally and inevitably been partitioned among the peoples that matter, and those that matter have had most say in the partitioning' (1924, p.281). Like many others of his time, Fairgrieve was influenced by environmental determinism, which matched well Social Darwinist thinking:

'It is now evident how the general course of history...has been controlled by geographical conditions...(p.269). Here, then, are the great geographical factors which have influenced man, civilised, semi-civilised, barbarian and savage....The great equatorial forest is no place where civilisation may grow...(p.274).'

There was a hierarchy of status from slightly to greatly unfavoured nations. An underlying environmental determinism was widespread in the way their homeland influenced their character. Even the Dutch were not immune from unfavourable comment. As a columnist in *The Teachers' Aid* argued:

'The overlooking of this affinity between the physical features of a country and the temperament of its inhabitants has been one of the most serious defects in our geography lessons....What...of the Hollanders and the manner in which they have been influenced by their surroundings; the prosaic nature of the scenery and the consequent lack of imagination of its people; their stolid indifference and dogged pertinacity which is so characteristic of the Dutch.' (1902a, p.355)

It may be argued that applying back current judgements on racism to earlier writers is anachronistic. Such a view can easily be countered, however, by the fact that there were geographers and educationists at the time who rejected the negative stereotypes being promoted, increasingly so in the inter-war years, as a new citizenship education supported by the League of Nations spread. Among geographers Roxby, and among geographical educationists, Welpton, expressed the need for developing sympathetic imagination and understanding of other peoples. Welpton as early as 1914 argued for a social geography that presented the world not as grasped by a geographer, but as grasped by a citizen, and sought understanding not only of physical conditions, but more of the lives of people on the earth. Unstead, author of a textbook series entitled *Citizen of the World*, took the view that children should be encouraged to be, in modern parlance, empathetic to other peoples: 'to try to imagine themselves in the place of other people...to think how they would act and live if they were in that environment, to see things with their eyes, to look at the world from their point of view...' (1928, p.321). Similarly, a Welsh teacher, Celia Evans, objected to the amount of imperial geography in textbooks and the emphasis on 'possessions' (1933) while others, like two Sudanese teachers, protested against the prevalent exotic stereotypes of primitive peoples:

'Teachers who have vivid imaginations, but little knowledge of the facts, find it easy to interest classes by telling exaggerated stories of the strange customs of savages....They emphasise the strange things in other people's lives and ignore what is similar to our own....' (Griffiths and Rahman ali Taha, 1939, pp.11–12)

While the more chauvinistic utterances of the late Victorian and Edwardian high tide of imperialism gradually faded, the imperialist spirit was still evident in post World-War II publications, though increasingly under the guise of 'Commonwealth Geography'. The Geographical Association promoted these ideas through annual conferences and in its journal. Laidler reminded teachers in 1946 that they must not forget 'the rich gift of Empire'. Similarly Howarth in his Presidential address to the Geographical Association in 1954:

'It may be said that I am inviting the schools to create a line of good little imperialists. And why not? ...Imperial geography must be the foundation stone of imperial citizenship...the more truly these stones are laid, the less occasion, in the long run, for racial misunderstanding and segregation.' (p.12)

The peoples ruled over by this Empire were presented as fortunate, their children as happy, with adult labour working contentedly in plantations in the warm sun, with the advantage of having lesser needs than their European masters. Spink and Brady accepted there were difficulties in race relations, but argued that not all were insoluble (1958). Relations between Europeans and Africans they saw as symptomatic of the mid-twentieth century issues at stake:

'In those parts where the European makes his home in Africa, it is especially serious...(Here) both races live side by side. Both are entitled to claim living standards suitable to their needs, but they are so vastly different in standards of civilisation that the needs of the African people are far fewer than the needs of the Europeans, and the African workers are able to live comfortably on much lower wages than the Europeans.' (p.62)

This particular defence of inequality had been more than matched in J.F. Houston's so-called *One Approach Geography–History* series, published in 1952:

'While Boer and Britain have gradually drawn closer together, the white and native races must remain separate...Black and coloured people, therefore, do not mix with white people and it is hard to see how they can ever do so. If the former were trained to the same level and allowed to do the same work as white people, earn the same salaries, and purchase land and property, they would in course of time, because of their greater numbers,. swamp the white population altogether. If that were allowed to happen, the civilisation which the white people have brought to South Africa might disappear altogether and the country sink back to the barbarism and tribal warfare from which Europeans have rescued it.' (pp.77–8)

As Marsden (1976b), Hicks (1981), and Wright (1983) have shown, stereotyping in textbooks is still with us, though more extreme imperialist and racist views are generally avoided.

Gender Stereotyping

Other elements than race, such as gender stereotypes, have also long been present. As part of the overall deterministic stereotyping of national characteristics, there were sometimes sub-sets, as in differences between males and females, in terms of appearance, temperament and roles.

In materials provided mostly for the primary phase, children's geography often loomed large, and the different roles of boys and girls carefully outlined, particularly in primitive societies. In a lesson plan in *The Teachers' Times* (1904) on 'The Kaffir Children' of South Africa, for example, it was stressed that in their uncivilised state the children grew up under the influence of nature. In their childhood play, the boys learned how to make weapons and hunt and fish, while the girls fetched water, which they 'carry gracefully in jars upon their heads.' The girls were also taught how to weave baskets, to till the sweet potato and maize fields, and make mealie porridge. On reaching marriageable age, the Kaffir girl would be sought by a man in a Kaffir regiment, or one with cows enough to buy her. She needed to walk for miles to see him in the preliminaries before marriage. These activities were so tiring that as night fell and the round huts were deep in shadow, 'the Kaffir lass creeps in through the small doorway, and...soon drops into a dreamless sleep' (p.129).

Generally, the children of primitive peoples were presented in a favourable light, and shown to be happy in ways experienced by children all over the world. The children of Japan were the happiest of all, not least because the filial respect, reverence and unquestioning obedience they showed to the parents, in return for which they were surrounded by parental kindness and loving care. The little girls needed to be hyper-sensitive to accepted social conventions. They were taught to refer to the lunches they prepared as unworthy and the house they had left as dirty, when they really meant the food was delicious and the house spotlessly clean. It would have been impolite to make less modest claims (McDonald and Dalrymple, 1910, pp.83–4).

In warmer climatic regimes, the women, as they grew older, were said to deteriorate more quickly in appearance, as reflected in lesson notes offered in an educational journal, *The Schoolmistress* (1918), giving background information to teachers and student teachers about Spain.

> 'Boys and girls have all the national characteristics. They are merry, cruel and lazy if they can be. They are certainly attractive. The dark hair and flashing black eyes are in contrast to the olive face. The girls are very handsome when young, but attain maturity much more quickly than their sisters in the northern countries and lose their looks long before an English-woman does. The middle-aged women look old, the old very ugly. Girls are, probably on account of their impulsive hot nature, kept more in seclusion and indolence than any other women except those in the East. The lover gets little chance to speak to his lady; he has to content himself with serenading her under the balconies which are such a common feature of the houses.' (p.116)

On the whole, gender stereotypes were less central in geography texts than those racial ones associated with the promotion of imperialism. They were more important, on the one hand, in determining different curriculum areas for boys and girls, with domestic economy looming extremely large for girls, and in literature, not least in the burgeoning magazine and novel literature for boys and girls. Many of the stories were set in exotic geographical locations. In the more melodramatic, the typical environment was that of inhospitable jungle or desert, inhabited by hostile natives and dangerous animals. The characteristic role of the male was to defend the females in his party not only from death, but from fates worse than death or, as they were more frequently described, 'fates that could not be mentioned'. A not untypical example can be taken from Kingston's *In the Wilds of Africa*, in which Stanley and David are fearful for their companions Kate and Bella as they canoe through hostile savanna territory:

'We now saw a large body of black warriors shaking their spears, and beating them against their shields, as they rushed towards the bank of the river... "Let me now take the paddle, Kate", said David... "My arm is stronger than yours, my sister, and in case the savages attack us, you and Bella must lie down in the bottom of the canoe". The canoes glided rapidly down the stream.... Had we not had the two helpless girls to protect, the adventure would have been an exciting one....'

Having passed through the danger zone, David makes clear his admiration for his sister:

'I had often read of heroines; but as I looked at the calm countenance of Kate, showing that she was resolved to go through all danger without flinching, I could not help thinking she deserved especially to be ranked as one'.

The hostile tribes were hardly to be distinguished from the animal life:

'I could see as I gazed over the plain, besides the negro army, numerous animals scampering across it, put to flight by their appearance....' (1871, pp.172–9)

Gender stereotypes continue to pervade geography textbooks, if less blatantly than those of the past. They are particularly evident in primary texts, where husbands continue to be portrayed as those who go out to work and their wives as those who stay at home, cook, wash and look after the children. Wright has explored sexist stereotyping, partly on the basis of inclusion (mostly males on photographs in work roles) and exclusion (women rarely represented in photographs, for example) (1985b). Women were rarely shown in active, enterprising roles, rather as passive and domestic.

The gender issue needs to be more fundamentally explored in the context of feminist geography. Though, as noted in Chapter 2, there are a variety of feminist perspectives, the fundamental clash would appear to be between integrationist and the radical feminist ideology (Williamson-Fien, 1988, pp.104–11). The integrationist perspective involves 'putting women in', making them more prominent in the landscape, and suggests that justice for women can be achieved through improving the nature of geography, and the way women are represented in its writings. A radical feminist perspective rejects this and argues that geography as such has only ever provided a partial view of the world, and is inescapably patriarchical and capitalist in its spirit and purposes. A few marginal changes to curricula can ease male consciences in the hope they will dispose of the sexism issue, but achieve nothing of substance in terms of the improvement of the lot of women.

In the radical feminist argument, justice can only be achieved by overturning a number of status quos, not least existing definitions of the structure of knowledge, of which geography is a part. Thus geography has no primacy in this equation, though its insights and techniques may tactically be permeated with advantage into a genuinely feminist critique. Radical change involves the introduction of a feminist pedagogy, in which female pupils are taken seriously and given equal opportunities in classroom discussion, for example, and in which teaching strategies revolve round the collection, analysis and assessment of data on women. A recent example on 'gendering space' is offered by Madge (1994). Patriarchal definitions of work and leisure are among the many areas of content which need to be recast. A reorientation of curriculum is therefore necessary, with subject frameworks seen as disposable and, presumably, areas of experience frameworks seen as more desirable. The radical feminist critique therefore categorically does not favour an approach, even in which human rights issues are put up front, which supports an issues-permeation into subject frameworks, and demands a particular issues orientation into which geographical frameworks, when regarded as useful, can be permeated.

Ageism

There is less vocal support for another highly disadvantaged group, the old, a neglected cohort in the world of geography, not least school geography (Marsden, 1988). Virtually the only mention of old age to be found in traditional geography textbooks has been in sections on physical geography, where river systems were typically divide into youth, maturity and old age. Carter's *Landforms and Life* (1961) offered some lively metaphors: 'the impetuosity of youth', 'the surging vigour of manhood' and 'the uncertain shuffle of old age'. Roget's *Thesaurus* (Section 131) strays more dismayingly into stereotypes of old age: '...grey hairs, senescence, senility, dotage...wrinkled, lined, rheumy-eyed, toothless, palsied, drivelling, doddering...past it...'

Apart from humanitarian, there are also pragmatic reasons for paying more attention to the geography of retirement and old age. Between 1981 and 2001, the total number of persons aged over 75 in Britain will have grown by 28%, to about 900,000 people. Old people form the most extended of all age categories, from 60+ to over 100. They cut across gender, ethnic and social groups. They are distinguished not only by age but also by inactive employment status. Old age is often equated with pension status. Law and Warnes (1976, p.453) conclude it is sensible to define the lower limit of old age as 60. The 60–75 group are not infrequently referred to as elderly, and the over 75s as very old. But these are aggregates, and at the individual level there is a wide spectrum ranging from the highly active to the heavily dependent.

In taking the geography of ageing and of old people seriously, such work should first be extended into the broader context of human rights education. The same principles that apply to multi-cultural and gender dimensions apply here. Old age is an issue that needs to be permeated either within geography, or in a cross-curricular framework. School textbooks remain a particularly inadequate source of information, and appropriate materials must be sought elsewhere. Among these are:

- Age Concern
  Bernard Sunley House
  Pitcairn Road
  Mitcham
  London

- Centre for Policy on Ageing
  25–31 Ironmonger Road
  London
  EC1V 3PQ

- Help the Aged
  Research and Education Division
  PO Box 460
  St. James Walk
  London
  EC1R OBE

Among many topics recommended for study, a significant number contain a geographical dimension. These include life-styles in old age; leisure in retirement; residence; access; and a range of social welfare issues.

Lessons for Today: a Case Study of Stereotypes of Spain

It is important to pursue the fundamental issue of social stereotyping today, even though such stereotyping may be less blatant than in the past. One reason is that there is plenty of evidence to suggest that government ministers once more see geography and history as patriotic subjects, and argue strongly for a bias as between British and world topics in favour of the former. They have attempted strongly to revive the spirit of national rather than international citizenship, a point to be returned to in the final section of this book. In these circumstances a more detailed and substantiated critique of nationalist geographies and their associated stereotypes is needed. This will demonstrate further that it is not only the social and political undercarriages of education, but also the subject paradigms and pedagogic principles, that beget stereotypes. The argument will be illustrated by a case study of stereotypes of a fellow-member of the European Union, Spain, exposing the way it and its people have been treated over the last 150 years, in four different geographical paradigms: for sure stereotyping did not die out with the capes and bays paradigm.

Capes and Bays and Catechetical Teaching

One of the pedagogical devices employed in catechetical teaching was what was later to be called by Bruner 'the power of contrast':

> '...by getting a child to explore contrasts he is more likely to organise his knowledge in a fashion that helps discovery...its efficacy stems from the fact that a concept requires for its definition a choice of a negative case.' (1966, pp.93–4)

Undoubtedly there is a strong instructional case for seeing children as learning things better if exemplars are plotted against non-exemplars, if differences between cases are emphasised. Strikingly, this principle was advanced not only by educational psychologists of this century, but was also central to progressive Froebelian pedagogy. In a series of articles on Froebelian theory, *The Teachers' Times* declared very forcibly that things became known only when connected with the opposite of the same kind: through 'the connection of contrasts' (1904, p.304). What the Froebelian was preoccupied with was moral opposites: rudeness and kindness, good and evil, and so on. But the power of contrast was equally to be harnessed in the politicisation of geography and history curricula.

One of the early nineteenth-century geography textbook writers, the Reverend Goldsmith (1813) was already implementing this procedure as a means of reinforcement:

> 'The most despotic and tyrannical governments are those of Morocco, Turkey, Russia and Persia. The freest governments are those of the United States of America, and England.
> 'In colonies, England has the greatest number, and Spain the greatest extent....The most civilised and philosophical quarter of the world is Europe: the most barbarous is Africa....' (p. 61)

A similar approach is vividly illustrated in Tate's student text *The Philosophy of Education: or the Principles and Practice of Teaching* (1860), in which England and Spain were presented as opposite poles in terms of physical and economic geography, and national character. Both locational information and value statements were offered as indisputable fact, to be memorised:

England	*Spain*
Forms the greater part of an island...	Forms the chief portion of a peninsula...
The climate is damp and changeable...	The climate is generally warm and salubrious...
Rich in coal...	No coal...
The religion is Protestantism	The religion is Romanism
Has advanced very rapidly	Has retrograded since the

since the Reformation...	period of the Reformation...
The workshop of the world...	Cannot supply its own people with manufactured goods...
The greatest country in the world...	One of the most contemptible states in civilised Europe...
The people are pious, industrious, generous and brave...	The people are bigoted, indolent, treacherous and base...
Its colonies flourish in every part of the globe...	Its colonies are dismembered and enfeebled...(p.191)

The Stereotypes of Deterministic Regional Geography

In the later regional paradigm, the Spanish people became stereotyped as a territorial and racial fringe group living in a climate intermediate between the cool temperate climates, seen as stimulating to Anglo-Saxon human energy and initiative, and creating great civilisations, and tropical regimes, encouraging the laziness and excessive sexual appetites of inferior Moorish and negroid types. Many textbooks told children that 'Africa begins at the Pyrenees'. Things became much worse when Africa really began, as in Morocco, a 'monument of barbarism', as described, nearly a century after Goldsmith, in *The Teachers' Times*, which presented itself as a progressive journal:

> '...the Moors are mentally, morally and physically a hopelessly degenerate race...Fez...like all the towns of the interior, is a mass of ruin and all-abiding filth...in a map of the world used in the geography classes England figures as a small island lying just south of Tibet...knives and forks are unknown, his Shereefian Majesty himself eating with his hands.' (1903, pp.178–9)

In earlier centuries, the Moors had conquered much of Spain and left an African taint. A range of cultural distortions was entrenched in the textbooks, in nationalist, racist and sexist stereotypes, and indeed in European high culture. Composers such as Mozart (*Don Giovanni*), Rossini (*The Barber of Seville*) and, perhaps above all Bizet (*Carmen*), set some of their greatest operas in Spain, presenting a rampaging set of stereotypes of macho men and sultry women, playing out rather primitive human relationships and intrigues in sub-tropical and emotional heat. Yoxall was predictably jaundiced in his summary of contemporary Spanish character in his *The Pupil Teacher's Geography* (1891), accepting uncritically all the handed-down caricatures of north European folklore. 'Sunburnt Spain', he recounted:

> '...is a State whose glorious past has decayed into a somewhat ignoble present...now but a third-rate European nation....The Spain of the present is the land of the bullfight, the siesta noon sleep, the bandit, the smuggler, the

corrupt official, the poor, proud, work-despising Don, the untaught, super-stitious, garlic-eating muleteer...' (p.176)

One of the exotic differences regularly alluded to in the texts was the bull-fight, a constant preoccupation of English travel and textbook writers. In Sandeman's contribution on the Iberian peninsula to the European volume of the *Highroads to Geography* series (1915), the opportunity is taken to contrast its colour and public ceremonial against the appalling cruelty and the crowd response:

> 'We shall not describe the revolting performance. It ends when the goaded, maddened animal is struck dead by the sword of the daring matador. The cruelty of the Spaniard's nature is clearly seen in his love of this debasing sport.' (p.228)

Pickles, whose texts classically illustrate the application of the regional paradigm to secondary school geography, approaches the issues more subtly, asking children to write down their images of Spain. These included the predictable: oranges, onions, guitars, black cloaks, bull-fights, duels, dancing, sunny skies and so on. But Pickles commented that 'no one thought of a man working'. Countering the normal stereotypes of his contemporaries, he went on to point out: 'We see so many pictures of Spaniards singing, dancing or idling in the sun, that we are apt to forget that they have to work hard like everybody else'. The nature of their work he claimed reflected the influences of geography, illustrated by the different regions of Spain (1932a, preface). The earlier stereotypes were of course those of the hot south, next door to Africa. It was conveniently forgotten that the north-west region of Spain was climatically and physically very similar to Britain. Interestingly, some texts did differentiate between the bulk of Spain and the Catalan region. Thus Pickles characterised the Catalan people as hard-working, energetic, determined and business-like, and Catalonia as for long the main centre of advanced ideas in Spain, the first to throw off the yoke of the Moors, and with a distinct language from the rest of Spain (1932b, p.186).

Case Study Stereotypes

As was noted in Chapter 3, the regional paradigm in school was from the 1950s partly overtaken by the more 'enlightened traditionalism' of the sample study or case study method. While the best of these studies were vivid and helped to bring the world into the classroom, they could, if not carefully introduced, infiltrate a new set of stereotypes. At least Pickles' regional geographies offered a reasonably balanced coverage of all parts of Iberia. How Spain was represented in the sample studies texts of the fifties and early sixties tended to be in the guise of 'oranges from Spain'. Those of the 1960s shifted very heavily into tourism and the Benidorm stereotype study, quite logically in terms of topicality, as more and more people from Britain went for holidays in Spain. But what do they know of Spain who only Benidorm know? Some of these case studies were of a high quality of presentation, but on the one hand offered a grossly attenuated impression of a varied country and, on the other, reinforced the exotic images of the tourist brochure.

European Studies Stereotypes

While geographers were at the time moving away from regional approaches, the trend towards integrated studies in the secondary school, in the wake of comprehensivisation, witnessed a return to area studies. These included a European Studies variant of social studies and, in some cases, offered links between modern languages and geography and, perhaps, history, often as alternatives for less able children not thought able to cope with straight modern languages courses. The course planning, aimed at topicality and stimulating interest, again tended to reinforce tourist stereotypes. Without a strong and distinctive disciplinary input, there was a falling back on topics such as wine and cheese, Paris shops and fashions, and so on, in the case of France. In similar approaches to Spain, there was a revival of old stereotypes, including the mantilla and the bull-fight (which had disappeared from geography texts and tourist brochures), the fruit-picker and the dancing girl, and perhaps some reference to heritage cities in the interior and the Moorish south, such as Seville and Granada, intended to give an overall 'flavour' of Spain. The World Cup of 1982 might have helped football to replace bull-fighting as the key sport in the culture of Spain, and certainly the entry into the European Community and the 1992 Olympic Games in Barcelona have assisted in updating the images.

A New Stereotyping?

As we have seen, pedagogic factors have shared with others responsibility for promoting negative stereotypes of other peoples. A latter-day version of this can be found in the cartoon strip. A noticeable trend in recent years has been the increasing use of cartoons in geography textbooks. Cartoons are in their strict sense satirical drawings which crystallise current thinking about famous figures, especially political, in a humorous and debunking way. They are potent influences in that they have high entertainment value and also help to simplify the real world. They provoke, polarise and confront. They are not concerned with educational principles such as building on a sound basis of fact, sifting evidence, refining concepts and exploring shades of meaning. They pack a simple message in a powerful punch.

Sophisticated political cartoons must be distinguished from comic-strips, which emerged in specialised publications for children and youth during the nineteenth century. But both are engaged in the process of caricature. The perceived advantages of using cartoons and comic-strip material in geography textbooks appear to have been:

- *motivational*, in providing entertainment value, designed to appeal in particular to young and to less able children;
- *pedagogical*, as a means of simplifying the complexities of reality;
- *presentational*, in the understandable attempt to offer a wider variety of stimulus materials;
- *logistical*, as a convenient strategy for making available generalised visual impressions of life in other places, in the absence of access to real photographs;

Figure 8.2 Inter-dependence? Hoisting the British flag in New Guinea

- *cognitive*, in the strict sense of using relatively sophisticated cartoons as teaching material, requiring the application of interpretational and problem-solving skills to the issues sub-text of the cartoon.

It will be clear that in any issues-based geography (*see* Chapter 9), the use of carefully chosen cartoons can readily be justified. Thus the late nineteenth-century cartoon-type print shown in Figure 8.2, intended to excite patriotic responses in its own time, contains a whole set of sub-texts of relevance to values education, which can be exploited in enquiry-based and problem-solving situations. It is emphatically not about this type of cartoon that criticism is being made.

Whether comic-strip caricatures and cartoon-type presentations in the format of talking-heads tied to speech balloons are equally justifiable can at least be questioned. Thus 'Beryl-the-Peril'-type caricatures babbling forth over-simplified pieces of geographical information or, worse, predigested opinions on serious world issues, usually far distant from the pupil's own experience, are incongruous, trite and, far from adding a human touch, are dehumanising. In some ways, it is worse than trivial pursuits, capes and bays geography. It is, hopefully, significant that many early reading books for children are moving away from comic-type characters to photographs of real children, events and places. One of the most galling aspects of comic-strip geography is its lack of attention to the distinctiveness of the subject, which is about real locations. Most of the activities depicted could be taking place almost anywhere.

Exercises based on such materials reflect a deficit view, less able children being given tasks based on comic-strip stimulus materials, and the more able being trusted with more conventional materials. Comic strips have tended to be linked with issues-based approaches, perhaps not surprising in that such approaches are less discretely geographical than those tied to place. The relative prevalence of these materials has, it can be argued, reduced the distinctively geographical nature of many secondary textbooks. The distinctiveness or otherwise of issues-based work is the focus of the next chapter.

CHAPTER 9

The Values and Attitudes Dimension: Issues-based Geography

Values and Geography

In Chapter 1, the central importance of a values dimension in the process of formulating educational aims was identified. Much evidence has been presented in other chapters to reinforce the conclusion that as a subject fundamentally involved with the global dimension, geography must take values issues seriously (Wiegand, 1986). In an interesting discussion of the application of the Kratwohl–Bloom taxonomy of educational objectives (1964) to values education, the so-called **affective domain**, Carswell (1970) suggested that before geography could deliver its contribution to values education, it had to instil in pupils some commitment to itself as a discipline. Carswell's interpretation remains worth noting, if it is in some respects contrived and over-classified:

1.0 Receiving:
1.1 Awareness – Awareness of geography as a discipline.
1.2 Willingness to receive – Acceptance of the importance of geography.
1.3 Controlled or selected attention – Sensitivity to news about geography.

2.0 Responding:
2.1 Acquiescence in responding – Willingness to read the assigned literature on geography.
2.2 Willingness to respond – Voluntary acquaintance with the current geographical literature (i.e. intrinsic motivation).
2.3 Satisfaction in response – Enjoyment of arguing current issues in geography.

3.0 Valuing:
3.1 Acceptance of a value – Feeling oneself to be a geographer.

3.2 Preference for a value – Assumption of an active role in geography.
3.3 Commitment – Loyalty to geography.

4.0 Organisation:
4.1 Conceptualisation of a value – Development of a rationale about the role of geography (e.g. do you agree that the role of geography is to help people better to understand the world in which they live?).
4.2 Organisation of a value system – Formation of judgements reflecting beliefs that geography provides a system of enquiry that helps to understand and solve world problems (e.g. do you agree that consciousness of world problems acquired through study of geography contributes to intelligent decision making?).

5.0 Characterisation by a value or a value complex:
5.1 Generalised set (learner ordering things around him in stable frame of reference) – Viewing of problems primarily from a geographic point of view.
5.2 Characterisation (formulation of personal code of conduct or philosophy) – development of a philosophy of life consistent with geographic theory and practices.

Despite its limitations, Carswell's taxonomy helpfully reminds us that:

- values education is complex;
- while values education is generic, there are specific interpretations within and distinctive contributions from particular disciplines; and there is a cognitive underpinning to the development of values.

The Liverpool Schools Council Project History, Geography and Social Science, 8–13 (Blyth et al., 1972) also identified pieces of evidence which revealed whether changes in 'attitudes, values and interests' could be justified as changes in the right direction, and hierarchically. In terms of interests, for example, distinction was made between:

- the child 'who shows a passive interest in people and physical features of an environment' but no more;
- the one 'who considers, willingly, questions asked by others about the environment'; and
- the one 'who spontaneously collects materials, or has a strong desire to find out things for himself about the environment', and so on.

There was similarly attention given to contacts between the affective and cognitive domains; and between the theoretical and speculative arena of arm-chair discussion and that of practical action. Thus, among the 'higher level' group of qualities identified were those of the pupil 'who is wary of over-commitment to one framework of explanation and is alert to the possible distortion of facts and omission of evidence', and secondly of the pupil 'who is willing to identify with particular attitudes and values about the environment and relates these to other peoples' (pp.5–6).

Issues-based Geography

During the 1960s there was a convergence in the respective thinking of the conceptual revolutions in curriculum theory (*see* Chapter 5) and the 'new geography' (*see* Chapter 2). The latter, as Alice Garnett had predicted, provoked fragmentation. One reason for the later splintering was that the 'new geography' was diffused by some of its disciples in universities and schools with an excess of zeal, as the 'one best system'. It is significant that at about the same time there were other well-documented educational trends, including advocacy of the progressive principles of the Plowden report, a perceived decline in standards associated with the expansion of comprehensive schooling and the ending of the 11+ examination, and student unrest in the universities, all exciting critical public debate. The combination of what was viewed as an accumulation of permissiveness, at all educational phases, was too much for far-right opinion to stomach. The long-term effect of this reaction, as previously indicated, was a developing polarisation and politicisation of discourse about education. This was evident in the field of geography, with one platform advocating a more activist, politically conscious welfare geography, while the other sought to return to the safer sea-lanes of value-neutral information dissemination.

The attitudes of children were increasingly regarded by the former group as requiring equal attention as learning capacities. One obvious argument was that positive attitudes promoted academic achievement. The other dimension, represented in the Carswell taxonomy, was that more consideration also needed to be given to the promotion of positive attitudes towards the environment and other peoples, as part of an issues-based framework.

One of the consequences of the social and environmental determinism and political jingoism found in both secondary and primary texts up to the 1960s was an increasingly sceptical reaction, not least in primary schools, as to the value of the subject in the curriculum. Was such geography worthwhile? Was not an integrated thematic framework a better option? The amount of distant place geography taught in primary schools would seem to have declined during the 1960s and 1970s. The malign consequence of this was that in so changing many schools effectively cocooned their children in the parochialism of local studies. Much of the early broader geographical entitlement was lost, at a time when research by Jahoda (1963) and Carnie (1966) was suggesting that children must be caught young if favourable attitudes to other peoples were to be developed.

In secondary schools the situation at the turn of the 1960s and 1970s was rather different, even though it was witnessing some shift away from geography as a separately timetabled subject. Interest in economically developing countries, their exploitation and the quality of life of their people, was predictably not confined to geographical education, for indeed the associated issues were and are multi-disciplinary. It was argued that a balanced picture of problem issues could better be achieved through integrated studies than through separately timetabled subjects. Additionally, the comprehensivisation of secondary schools in the 1960s and 1970s also led to calls for a shift from a subject-centred (seen as more suitable for the academic mind) to an integrated structure in the secondary curriculum (seen as more conducive to motivating less-able pupils). Thus many schools switched in the lower secondary classes from separate courses in history and geography to integrated humanities or social studies.

As we have seen (*see* Chapter 2), there was among geographers at this time some revulsion against the abstraction, dehumanisation, and retreat from social relevance that the positivism of the quantitative revolution was seen to represent. In turn new foci emerged, including behavioural geography, environmental geography, welfare geography and the like. These various labels can roughly be collapsed into two main thrusts: **humanistic geography** and **radical geography**.

Behavioural geography was perhaps an unsatisfactory title, in savouring of behaviourism, with all its positivist connotations. Phenomenological thinking moved us into a more **humanistic geography**, in which key features were:

- the identification of human agency and individual perceptions as geographical factors in themselves;
- opposition to positivism and environmental and social determinism;
- a shift from the objective to the subjective;
- a focus on the individual and the real rather than the aggregate and the abstract;
- a suspicion of ideologies as closed systems of thought.

In comparison with more radical variants of welfare geography, even though humanistic geography stressed a conscientious engagement with social issues, humanistic geography could be seen as a more disinterested academic pursuit, 'based more on ideas than ideologies' (McKewan, 1986, p.161, quoting Buttimer).

In school terms, McKewan maintained that adopting a phenomenological approach to the curriculum demanded that geographers should present their subject pedagogically in a manner consonant with that view of the world, which necessarily would involve seeing students as active, would question taken-for-granted assumptions about the curriculum, developing it from the experience of those students, raising their consciousness through subject matter related to their daily lives, and attacking issues, without the imposition of ideological parameters (pp.164–5).

The other heavily over-lapping paradigms were different not only in prioritising controversial social and environmental content, but also in encouraging a commitment actually to do something about the issues: in other words, to engage in social/political action. It might be argued also that they were characteristically located within a more constraining ideological framework.

In school terms, the developing environmental concern of the 1960s was reflected in a shift of emphasis from environmental study (the environment for education) to environmental education (education for the environment) (Martin and Wheeler, 1975). Pioneering attempts at producing teaching packages reflecting the geography of social concern, as represented in the GYSL materials, could similarly be regarded as **social concerns for education** rather than **education for social concerns** – in that studies of decaying urban environments, leisure activities, and the like were seen first and foremost as a means of catching the interest of less able children, and therefore of being especially relevant to the needs of inner city schools and young school leavers.

The distinction between social concerns for education and education for social concerns was evident in varying level of commitment to the coverage of controversial social and environmental issues in secondary school geography. Different intensities of issues-pervasion into the curriculum had become identifiable by the early 1980s, namely:

- *issues-permeated* geography,
- *issues-based* geography, and
- *issues-dominated* geography.

Returning to the 'new paradigms', it can be argued that environmental and social welfare geography led us at least into the issues-based mode, while the radical branch of welfare geography demanded issues-dominated work. One of the problems of the latter preoccupation was that it could, in over-enthusiastic hands, create undifferentiated and pathological stereotypes of the global village. A more important problem of principle is that issues-based approaches are by definition multi-disciplinary. For if the prime focus is on the issue, then disciplines like geography, but never geography alone, must permeate and illuminate the themes. This was the basis, for example, of the later world studies courses, so effectively delineated in Fisher and Hicks (1985). They offered an entirely legitimate way to proceed, though not the only way, and not a distinctively geographical way. Fisher and Hicks accepted the value of geographical permeation into a world studies approach, rather than the alternative of a global geography into which issues were permeated. Clearly the import of this book is that issues-permeated geography, properly applied, has some advantages, to be argued below.

Various socially-oriented forms of integrated study in the 1970s and 1980s replaced earlier variants, including a number strongly related to welfare geography, such as environmental education, world studies, global education, peace studies, and human rights education. Imaginative and committed teaching ideas emerged, for example, in books co-authored by Hicks and Steiner (1989), from Pike and Selby (1988) and in Selby's *Human Rights* (1987). An important methodological source was Huckle's *Geographical Education: Reflection and Action* (1983), as later was Fien and Gerber's *Teaching Geography for a Better World* (1988). The theoretical principles were supplemented by vivid first-hand materials concurrently being produced by world aid and other agencies, including the Centre for World Development Education in London and the Development Education Centre in Birmingham. Children from primary level upwards were rightly encouraged to show active as well as merely armchair interest by engaging in school links (Beddis and Mares, 1988), and fund-raising exercises in school.

There is, however, a difference in principle between what would widely be agreed to be laudable educational aims, and moving into an arena so dominated by issues that the self-evident good causes embraced appear to justify indoctrination. Right-wing forces asserted that initiatives such as human rights education and peace studies were a front for left-wing indoctrination, often on the basis of highly selected abstracts from the more radical left-wing literature. There were many more pursuing these initiatives from a just-left-of-centre position. But the prevailing polarised level of the political debate could not cope with the nuances. At the same time, it is harder to deny that enthusiastic proselytisation of welfare geography can lead to a shift from educational detachment. Subjects like geography and history, and various types of integrated studies, have undoubtedly been used in the past for the promotion of what were accepted in their time as undeniably good causes, whether to do with religion, imperialism, eugenics and health education, or capitalism. We will return to this important issue in the context of the cross-curricular matters associated with the National Curriculum (*see* Chapter 10).

Most of the methodological material on peace studies and human rights has been careful to stress the educational orientation of the work, presenting both sides of the particular case, though at the same time appropriately suggesting that passive thinking, still less a diet of factual information, is not enough if we were arguing for education for a better world. The 'reds under the bed' scare-mongering of the early 1980s was, however, enough to generate a 1986 Education Act in which one of the clauses referred to education in political matters, which forbade LEAs and governors to allow the pursuit of 'partisan political activities...and the promotion of partisan political views'. Where political matters were introduced they should be in the form of a balanced presentation of opposing views, with pupils encouraged to form their own views on the basis of evidence. These latter points would be widely accepted by teachers of different political hues. The suspicion that this was the acceptable face of a hidden agenda was confirmed in the nationalistic nature of certain of the subject Orders and in the cross-curricular guidance (*see* Chapter 10).

While dismissing the phobias of the right wing as panic-mongering, there may be legitimate concern that the social education element has became top-heavy and tended to produce distortion, through disturbing the balance of good practice as between geographical, educational and social variables. It has already been pointed out that geographical distinctiveness has in many cases been lost in issues-dominated curriculum materials. The careful textual build-up of a balanced selection of evidence has in some cases been overridden by slogan-type and for/against generalisations. Whereas the materials of, for example, the *Geography 16–19* project, were packed with potential evidence as support for enquiry-based learning (Naish, 1985), by contrast, the so-called enquiry approaches promoted in texts for younger and less able children appeared in some cases little more than an incitement to offer an opinion.

In educational terms, far from generating meaningful learning and reflective thinking, which are surely among the key criteria of educational worthwhileness, texts and internally produced worksheets have fostered simplistic responses of young people to profound and intractable world problems, as Michael Storm (1983) discerned: '…we are asking them to build generalisations and to analyse complex relationships on the basis of a very slender stock of basic information' (p.38). One problem is that while the pioneers of such approaches have usually been rigorous in demanding a cognitive underpinning for work in the values dimension, the disciples have become, as with past paradigms, over-enthusiastic and over-simplistic in their presentation of issues. But it is falling into the trap of the ideologue and the extremist to build an argument on the worst rather than the best practice of opponents. The discussion which follows is presented not in polarised terms, therefore, but as offering a constructive critique of alternative and, perhaps, complementary approaches.

World Studies or Global Geography?

Let us contrast strengths and weaknesses of a world studies-type issues-based or issues-dominated approach, and those of an issues-permeated scheme implemented through a distinctively geographical framework. One of the most valuable and inspirational sources for the former is the previously cited Fisher and Hicks *World Studies* handbook. Apart from its helpful and imaginative practical ideas, its key characteristics are as follows:

1. *A non-confrontational pedagogic approach* – accepting the potentially valuable contribution of standard subjects in the curriculum to world studies (pp.22–3), by way of contrast to the polarised assertions of some promoters of integrated studies.

2. *A balanced set of objectives* – covering knowledge, attitudes and skills. In hindsight, it is a pity that, as in the National Curriculum, knowledge is separated from skills. Thus under knowledge it is specified that 'pupils should know how to investigate and reflect on a variety of possible futures' and under skills that 'pupils should be able to find out and record information'. It can be argued that the describing, explaining and evaluating that are tied to 'knowledge', are indeed also on the same wavelength as the enquiry and grasping of concepts that are consigned to the 'skills' dimension. It can be claimed that, if anything, the former are more like skills and the latter more like knowledge and understanding (pp.25–6). This is not a pedantic point in that, as we shall see later in the National Curriculum discussions, knowledge is falsely stereotyped as factual recall and skills as something more worthwhile.

3. *A check-list of concepts* for use in world studies. These are the broad generalisations of a type found in the earlier Liverpool Schools Council Project, including:
   – causes and consequences
   – communication
   – conflict
   – cooperation
   – distribution of power
   – fairness
   – interdependence
   – similarities and differences
   – social change
   – values and beliefs.

4. *A check-list of classroom topics* – such as 'the world in our newspapers', 'role-playing: everyday conflicts', 'detecting bias' and the like, which explore the above concepts.

5. *A check-list of key questions*, relating to the context of world situations:
   – What is the historical background?
   – How and why have things changed?
   – Who gains and who loses?
   – What conflicts of interest are there?
   – How fair is the situation today?
   – How are things likely to change in the future?
   – How ideally should they change?

6. *Check-lists for stereotyping*, including gender stereotypes, and those which relate to teaching about 'developing' countries and about minorities. The 'developing countries' check-list challenges teachers as to whether they are presenting:
   – a tourist-eye view (emphasising the exotic)
   – a packet-of-tea view (happy natives producing our food)
   – a pathological view (focusing on disasters and desperate poverty)

– a pat on the head view (follow our example and all will be well)
– a poverty as an act of God view.

Such materials are as important to the teacher of geography as of world studies. But why is the Fisher and Hicks blueprint a world studies and not a distinctively geographical approach?

The first difference is that the world studies approach as here represented conveys little or no sense of specific places. While references are made to moving out from the personal and community to world scales, detailed place study is very largely absent. Thus the themes suggested are, characteristically, 'women in the world', 'world population and wealth', 'the problem of tourism', 'why poor countries are poor', 'arming the world' and so on. Virtually all the maps included are world maps. There are almost as many cartoons as maps. The photographs used are not place-specific and not landscape-oriented.

As with old-time regional study, there are pedagogic problems in such a heavy focus on world distributions and the dominance of great world issues. This is in no way to argue that children aged 8–13 should not be introduced to such issues. But it is to say that there are strong counter-arguments, as recounted in Chapter 3, for bringing the world into the classroom through vivid and detailed studies of particular places. Going from the familiar to the unfamiliar, the particular to the general, and the near to the far, is a time-honoured and effective pedagogical approach. It is very difficult to accommodate to a procedure which jumps so quickly from the individual to the aggregate; and from the personal scale to the global. There is currently more in issues-based work about aggregate global human rights, than about individual human relationships, as played out in particular localities. It is easier to be strong on human rights violations as the blame can generally be attributed to other people. Promoting good human relations both at home and generating empathy for particular people in particular places further afield appeals as a more appropriate starting point, important though it is to foster broader human rights education as well.

Another point is that, as was explored in Chapter 2, recent work at the frontiers of geography and also in the social sciences, has demonstrated an increasing awareness of the overriding importance of particular place variables in approaching and understanding the world's social and economic disparities.

A second basic issue relates to the limitations of the use of generalised key concepts such as interdependence, similarities and differences, and the like. These provide valuable frameworks up to a point, but their level of generality is such that they cannot offer detailed advice on how to select content. Here the complementary notion of more specific **principles** or **key ideas** is more useful, and is one which has been widely used in the secondary geography Schools Council projects and many methodological and textbook materials. It was illustrated in the European Studies context in Chapter 4.

Thus in the world context, key ideas could be, just to take a small selection based on Barke and O'Hare (1984):

- **under-development** is not necessarily an initial state or condition, maintained by internal **resource deficiency**, but a process generated by the interaction between the less developed and the more developed countries;
- the key to third world under-development would seem to lie in the past, in **colonialism**;
- localised economies specialising in export cash cropping and mining were established in the colonial era and have become vulnerable to **fluctuations in market demand;**
- despite **political independence**, the less developed countries remain **economically and technologically dependent** on the more developed countries;
- these continuing national forms of dependency are referred to as **neo-colonialism**, etc. (pp.71–2).

Such key ideas, which could be replicated in different forms, are therefore over-arching principles which link important concepts, vital to the understanding of the issues involved, and offer structures for initiating schemes of work. Thus many of the methodological texts and articles about development education include extremely useful check-lists covering the social and affective dimensions, but sideline the importance of the educational and the cognitive domain, which should be based upon the conceptual frameworks of particular disciplines such as geography and economics. A strong conceptual structure is always a necessary basis.

A Case-study of Trans-national Corporations

How many, for example, have more than a stereotypical awareness of the nature of trans-national corporations? In many current views, like the caricatures of pygmies in old geography textbooks, they emerge as pretty nasty things, inextricably linked with the exploitation of less developed countries by more developed countries, and the potential destruction of the planet. Let us look for a more differentiated and nuanced view through a series of key questions.

What are Trans-national Corporations?

The geographical dictionary definition tells us that they are firms which manufacture goods or provide services in many countries while directing operations from a headquarters based in one country. The overseas plants are often established to:

- avoid tariff barriers imposed on imports into the countries concerned;
- take advantage of cheap labour, raw materials and energy.

Where are they Based?

Largely in countries of the 'capitalist' world, including North America, Western Europe and the Far East, though there are smaller trans-national companies in the less developed world tending to invest in even more under-developed states.

Where do they Establish their Operations?

Not only in less developed countries but also in more developed countries themselves. Examples are Japanese industries in Britain and other western capitalist countries. Much more foreign investment is directed into other developed countries than into under-developed. Since the overthrow of communist governments in eastern Europe, there has also been a vast increase in trans-national corporate interest in these countries.

What is their Economic Significance?

It is widely agreed that trans-national firms are the main vehicles for the internationalisation of capital which refers not just to money as such, but to capitalist economic relations of production. As we saw in Chapter 2, such internationalisation shows little respect for local employment and social concerns, and is characteristically preoccupied with short-term fluctuations and the need to respond to the quick-fire decision making of the world currency markets, rather than with longer term indigenous needs.

What is their Cultural Significance?

Again this is generally viewed negatively, with a global spread of international advertising agencies and consumerist media portrayed as reproducing western values at the expense of the local. On the other hand, individual people in poor environments of the underdeveloped world would appear to appreciate this provision and view satellite and other television, where available, with avidity, as a means of escape from the oppression of their every-day existence. It is also claimed that such internationalisation of news and documentary services is a genuine threat to restrictive totalitarian regimes in some developing countries, the population preferring to tune into satellite coverage than follow the interminable official party lines purveyed by the national television system.

What is their Political Significance?

The large trans-national corporations may have access to more wealth than the total GNP of many developing countries. This gives them enormous political power, and capacity to maximise possibilities of tax avoidance and avoid problems related to local environmental and other legislation.

What is their Environmental Significance?

This is also presented as generally negative and, certainly in the last analysis, economic priorities are more than likely to outweigh environmental.

Are they of any Benefit?

Most of the literature would suggest they are only of benefit to their share-holders. On the other hand, states and regions within states spend much time, energy and money on striving to attract foreign investment as a means of increasing domestic employment and wealth, as in the case of the Japanese car industry in Britain which, in economic terms, has been seen to be a trigger to improving economic efficiency in indigenous companies, as in car components.

While strongly inclined to the view that the economic, environmental, social and cultural influence of trans-national corporations is generally malign, the point must not be lost that:

- no more than the pygmies of old-fashioned texts should these commercial giants be over-simplified and stereotyped as look-alike;
- presentation of an undifferentiated aggregate 'bad guys' image can be regarded as symptomatic of the buck-passing propensities so character-istic of our age, implying that we cannot ourselves be held responsible for external problems, and that we can justifiably sit comfortably in our armchairs, practise minimalist morality (*see* Chapter 1), and encourage speculative classroom debates that someone else out there, too powerful for us to influence, is responsible for the world's social and economic evils.

In terms of moral education, there is also arguably a tendency for issues-dominated and even issues-based approaches to concentrate on aggregate human rights rather than individual human relations. An advantage of the more specifically geographical approach to these issues is that at a detailed place level teachers and students can get to grips with individual and group human relations that both enlighten and highlight the complexity of issues and promote personal empathy, which a head-on aggregate, generalised and non-place-specific human rights approach is perhaps less likely to be able to accomplish.

Permeating the Trans-national Corporation Issue into an Educational Experience

As illustrative of an **issues-permeated** approach, a perhaps provocative example of a trans-national firm, in this case the well-known fast-food chain, McDonald's, is offered. An attenuated version of such an approach submitted to the National Curriculum Council can be found in its publication *Geography and Economic and Industrial Understanding at Key Stages 3 and 4* (1992, pp.19–20). The aggregate cultural stereotype of such organisations is, as already indicated, almost entirely negative. In terms of health and environmental aspects, for example, the corporation is variously presented as rendering the streets ankle-deep in litter, lowering the quality of diets, and destroying the rain forest and the ozone layer. Yet, though decried by press, politicians and pedagogues, it is ubiquitously popular with the punters, whether in Manila or Moscow, Paris or Prague, Los Angeles or London. It testifies to the similarity of outlook in a population cohort of significant size the world over (if only perhaps in the propensity slavishly to follow the example of neighbours and the blandishments of the media), and demonstrate the power and universality of the large corporation in the global village. McDonald's is quick to deny that it ravages the rain forest and the atmosphere, and publishes copious evidence to the contrary. There are many interesting educational lines to follow in using the organisation as an issues-permeated geographical topic:

1. In general terms, what counts as evidence in environmental and health education? How can it be verified? Certainly either McDonald's or its critics are economical with the truth. Thus the firm claims that in any country almost all its raw materials come from indigenous sources, and none from raising cattle in the rain forest, and that its products promote healthy eating. Which is correct?

2. From the geographical point of view, the study addresses basic concepts of:
   (a) **Location** – Whether town centre or peripheral drive-in store, in a local study.
   (b) **Settlement hierarchies** – Which places do and do not have McDonald's?
   (c) **Spatial disparities in wealth** – Which continents and countries have fewest and most McDonald's?
   (d) **Spatial dissemination** – How have McDonald's spread over the world, how quickly, and why so successfully?
   (e) **Catchment areas** – Where do the stores on the urban periphery, for example, draw their customers from; and with which stores do they

compete for custom?

(f) **Cultural differences** – Why does Germany have over 350 McDonald's outlets and France, with a similar population and levels of wealth, less than 150?

3. At the cross-curricular level, there are obviously important links with economic and industrial understanding and, as we have seen, with health education and environmental education. In the case of EIU and environmental education, topics studied could include:

(a) concepts of **consumer preference** and their impact on production and distribution;

(b) changes in **consumption patterns** and their impact on industrial development;

(c) changes in the **location of retail services**;

(d) **regional and global economic similarities and differences** in the location of retail services;

(e) how **local planning policies** affect the location of retail outlets;

(f) how the production and use of some goods **damages the environment**;

(g) how producer and consumer **decisions** affect the environment;

(h) the impact of **trans-national companies** on the local economy.

The pupil activities could range from locational knowledge work to mapping distributions, from comparing these distributions in terms of population and wealth, to undertaking a simulation exercise to find the best site for a fast-food outlet on the edge of a great European city.

Such relatively detached approaches may be open to the charge of reflecting too neutral a stance and failing sufficiently to encourage radical action. But they at least explore complexity and maintain a balance between social, educational and geographical aims. One problem of more direct invitations to activism, and implicit if not explicit working on the consciences of young people, is that they can as readily produce negative as positive responses. In being invited to reply with pre-digested answers (so often met with in 'talking heads' illustrations of social and environmental issues), rather than to tease out issues for themselves by being given time to explore a wide range of information and opinion, students are being encouraged to leap a lot of steps in the tortuous path towards a conscientious and constructive personal understanding and empathy, let alone partaking in intervention.

The responsible geographical educationist must be true to the first educational calling which is both cognitive and affective: to promote knowledge, understanding and other thinking skills, in making children aware of issues that impact greatly and sometimes gravely on peoples round the world, and to introduce them to a set of values and attitudes founded on the humanistic and international ethics which underpin the universal declaration of human rights. If these high ideals were achieved then we would be taking steps to improve the quality of life and create a better world. But there remains an uncomfortable feeling that the manifest desirability of this cause is too easily translated into the inculcation of belief systems not entirely unlike those deployed in religious instruction of the nineteenth century, on the grounds that children were thus fitted for a better life hereafter.

The McDonald's example is a useful test case. It is a capitalist, transnational, fast-food, American chain: four key words in the demonology of many social and environmental educationists, and indeed of welfare geographers. However instinctively (and, in this case, possibly chauvinistically) hostile we might be to an organisation such as this, cannot still a valid distinction be made in the way we would wish to deal with the complex associated issues in the classroom and on political and media platforms?.

Games and Simulations

Over the last 25 years, games and simulations have consistently been regarded as a formative way of pursuing experiential, enquiry- and issues-based geographical activity. To an extent, the original introduction was associated with the Madingley model-building promotion of the late 1960s (Walford, 1969), but their potential for association with, first, behavioural geography (bringing in the dimension of human decision making) and, secondly, with issues-based welfare and environmental education approaches, were later equally appreciated. Games and simulations are also co-operative ventures and, at their best, promote personal and social education, and are motivating. If associated with vivid, first-hand materials related to actual places, they can be used to promote generalisation and empathy. On the other hand, if generalised, non-specific games are used to start with, they can produce a very detached and unreal view of major world issues.

Walford (1986) has summarised the regularly used themes for games and simulations in geography:

- location
- route building
- search for and exploration of mineral resources
- development of a land surface
- primary activities in the environment
- issues of environmental conflict (p.82).

As an example, a game linking tourism and conservation of the environment is offered in Figure 9.1.

As with other desirable activities, games and simulations offer no panaceas and need to be cautiously approached. They are time consuming and can become competitive pieces of entertainment, in which excitement about winning can mask the basic educational objectives being sought. The translations of game playing from the mores of the quantitative revolution to those of humanistic welfare geography clearly raise the stakes. For games and simulations are simplifications of reality and unless the simplifications are sensitively accomplished, they can result in distorted and even error-laden learning (Walford, 1969), and subvert the promotion of empathy that is fundamental to the purpose of the more attitude-based activities. That does not mean that such experiences should not play a central role in issues-permeated geography.

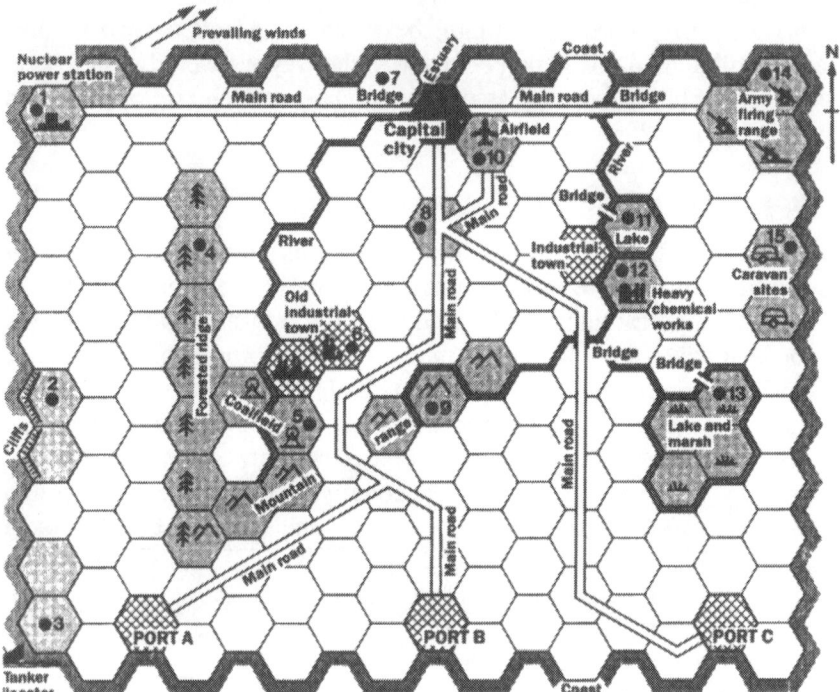

Noitullop is an imaginary island off the coast of Britain. Its king, whose family had ruled for generations, has been overthrown by the forces of the neighbouring island of **Neerg**. They find that the whole of Noitullop is in a very run-down state, with problems resulting from a waste of resources, and from a lack of care for the countryside and its wildlife. The new, conservation-minded ruler from Neerg asks United Nations' experts to come in and make a rapid survey of the conservation problems, so that measures for improvement may be taken.

How to play the game

1. Two (or three) players can take part. They start respectively from Ports A and B (and C). Each player needs a small counter. A coin should be tossed to decide who should start. Each player is a United Nations' expert.
2. Each player's aim is to get as quickly as possible to the capital city to make a report. Before doing so, he or she has to visit **ten** out of the fifteen conservation problem sites. These are marked by black dots and numbered.
3. Players move in turn, either straight or diagonally across the 'squares' (hexagons in fact).
4. Each player can move either (a) **up to three** 'squares' at one turn, or (b) **up to six** 'squares' on main roads. Players **cannot**, however, move on **both** main road 'squares' **and** non-main road 'squares' **in the same turn**.
5. On reaching one of the numbered sites, a player's turn comes to an end, even though he/she might have moved less than three squares.

Figure 9.1 Tourism: the Noitullop conservation game

6. As each site is visited, the player should mark down carefully on a separate sheet of paper the number of the site.
7. A player cannot count a visit to the same site more than once.
8. A player cannot move on to or over a 'square' on which another player is standing.
9. A player's turn comes to an end on reaching a river or lake, unless the 'square' has a bridge on it. The player can, however, move as usual during the next turn.
10. The winner of the game is the player who reaches the capital city first, having completed ten visits to problem sites.

Follow-up

1. When the game is finished, fill in the following chart. List the ten sites you visited, and beside each one mark in the correct conservation problem (from the list below) associated with the site.

Number of the site	Conservation problem	Number of the site	Conservation problem

Conservation problems (not in any particular order): **Sea-birds' nests; Radioactivity; Oil-polluted beaches; Rare trees; Breeding wildfowl; Derelict land; Atmospheric pollution; Traffic congestion; Noise pollution; Litter; Unexploded ammunition; Over-used ski slopes; Untreated sewage; Chemical pollution of water; Smells from factory.**

2. Select **four** quite different problem sites you visited, and suggest briefly a solution for each one.

(a) _____

_____

_____

(b) _____

_____

_____

(c) _____

_____

_____

(d) _____

Section D

NATIONAL CURRICULUM PLANNING IN GEOGRAPHY

CHAPTER 10

National Curriculum Geography in its Cross-curricular Context

Background

Supporters of whole curriculum planning have regretted what they regard as the demotion of cross-curricular activity in the National Curriculum to the status of a poor relation: a means of plugging the gaps left after the academic subjects and religious education have been found a place. Yet many of the flagship publications of the National Curriculum Council, including the Curriculum Guidance documents, were addressed to whole curriculum planning and cross-curricular dimensions and themes (*see* Hall, 1992). There was, therefore, face-value official support, at least from the National Curriculum Council.

There is little to suggest, however, either in these documents or in other discussions surrounding the build-up to the National Curriculum, that the cross-curricular themes chosen, from an army of contenders, have had a long history in the curriculum, though not necessarily under these labels. The fact that this history is in many respects a dubious one is either ignored or not known.

The purpose of this chapter is in part to bring to bear the historical evidence to support the view that, if we are concerned about the politicisation of the curriculum, there is perhaps more to be said for the permeation of cross-curricular issues into a subject framework, than for the permeation of subjects into a cross-curricular 'areas of experience' framework. One of the main reasons for this view relates to the hidden and even unhidden sub-texts of the cross-curricular dimensions and themes, as defined in the National Curriculum. These emerge strongly in the NCC Guidance documents, though it is evident that the Department for Education has at least for the present sidelined these statements, and is at least ambivalent in its attitudes towards cross-curricular activity. The chapter will demonstrate that the important potential contributions the cross-curricular themes have to make to whole curriculum planning have, historically, been misapplied by political lobbies, including the Church, in pursuit of their particular good causes (*see* Marsden, 1989). They have used the curriculum and related informal channels of education to serve extraneous ends, which essentially have involved instruction, control and 'conversion' of young people into passive and placid subjects, rather than the education of autonomous, reflective and critical-thinking citizens.

Instruction achieves its controlling grip through inculcation (the process of forcibly impressing on the mind by frequent repetition and admonition), and indoctrination (the associated content – the doctrine or belief system being inculcated). It will be clear from the many statements of Secretaries of State for education, other ministers, and in the right-wing press, that there is a either a very hazy concept of the distinction between education and instruction or, alternatively, a clear belief that the latter is preferable. It is also apparent that the educational agenda is very strongly politicised. As has already been pointed out, there is much evidence to support the view that the National Curriculum has been propelled in the direction of a nationalistic curriculum. This direction is not changed in the Dearing recommendations.

Instruction for Imperial Citizenship

Much cross-curricular activity in schools has historically been offered under the banner of citizenship education. In the first part of the nineteenth century, religious instruction pervaded the curriculum, and notions of good citizenship were closely tied to the idea of being a good Christian. The miseries of the current existence of servility and compliance for most of the populace were to be compensated by rewards in the life hereafter.

After the 1870 Act scriptural studies, while remaining important, had to face stiffer competition in the light of the overwhelming demands from respectable groups in the population for a more meritocratic, vocational and secular provision. In this context priority tended to shift to the training of good citizens for the present life. One of the most characteristic aspects of this change was the promotion of nationalism. Good citizens could ideally be defined in turn as patriotic subjects nationally, loyal imperialists globally, and morally as respectable church-goers in their own communities. Religion still suffused the more secular branches of the curriculum, however. For example, the successful development of an empire could be presented as evidence of divinely granted favoured nation status. Examples of the blatantly unfavourable stereotyping of other peoples can be found in Chapter 8, where it is made clear that geography and history, in the competition of subjects for a place in the curriculum, were beneficiaries of this change, in their capacity readily to accommodate within their provision patriotic and imperial sentiments. The key point was that for over a century a version of citizenship education openly pervaded geography and history teaching, and blatantly politicised their content.

Instruction for an Industrial Nation in Decline

Sarah Trimmer was a touchstone of educational thinking around the beginning of the nineteenth century. She was clear that educational opportunity should be offered to all the poor, to rescue them 'from that deplorable state of ignorance' in which they languished, but not to train them in any way that would 'raise their ideas above the very lowest occupations in life, and disqualify them for those servile offices, which must be filled by some members of the community' (1801, pp.22–3).

For Trimmer and later writers mass elementary provision meant 'the education of the poor', a concept lasting until the middle of the nineteenth century. By this time, as was noted in Chapter 7, it was evident that the growing and increasingly influential respectable upper working- and middle-class elements in the population were seeking training and accreditation in the skills required in the expanding public services, for work as schoolteachers, civil servants, and as clerks and typists in the burgeoning tertiary activities of large commercial towns and cities.

Concurrent with this trend was the emergence of the thesis that British industry was losing its pre-eminence, generating panics about the perceived inferiority of British entrepreneurship and the dangers of German competition. The finger of blame was pointed at the education system, accused of failing to produce enough skilled industrial workers. The notion of the 'English disease' had been implanted and was to become a fixed idea in the perceptions of a coming century of critics (*see* Roderick and Stephens, 1978).

Much later (1981) an American writer, Wiener, revived the thesis, claiming that the nation's industrial decline reflected innate British anti-industrialism and anti-urbanism, linked with a long-standing elitism in secondary and tertiary institutions, which had over a long period distanced themselves from the interests of industry and commerce. In an investigation of geography and history school textbooks, however, Ahier (1988) did not find this thesis substantiated in any direct sense. What he did infer was that the stereotypical presentation of Britain as the safe, benevolent homeland of a united race imprinted complacency and diverted attention from the values of industrial labour and commercial enterprise.

In general, it can be argued that the proselytisation of economic and industrial awareness, and of vocational teaching, has never been a disinterested educational endeavour, but rather a means of corralling a semi-skilled and unquestioning work-force in the good cause of meeting national needs. In the case of girls, the nineteenth-century priority, for the working classes in particular, was to produce prudent and thrifty housewives, to guarantee moral improvement of the coming generation. The metaphor of the thrifty housewife became in the 1980s a parable for the running of the nation! The recipients have, contrarily, been inclined rather to interpret the agenda as designed to fit them as cheap labour in unpopular and low-status occupations. The problem for British promoters of education for economic and industrial understanding is perhaps that too many children have either been educated enough or been street-wise enough to dig under the rhetoric. They might argue that it is for industry and government rather than schools to make the prospect of industrial employment more attractive and secure. This is in no way an argument against education for economic understanding: rather suspicion of linking it so explicitly with industrial promotion.

Health Education

As another theme long thought of as meriting pervasion across the curriculum, health education has an equally vigorous history. It is also one that provides some of the most lurid examples of indoctrination. Like other cross-curricular areas, its overlaps with moral and religious inculcation are strong. From an early stage, promoters of health instruction presented a negative stereotype of the subversive 'intellect-ualisation of the education of the lower orders'. Physicians in Britain and the USA attacked 'the vicious system of extreme study' and argued for a more vocational approach in schools, in which 'regular labour forms part of the system', an approach predictably geared to the interests of the middling and upper classes of society (King, 1838, p.183).

It was once more the social and moral perils associated with densely packed urban milieux that fuelled public anxiety. From the middle of the nineteenth century feverish responses to the perceived decline in urban racial health became a diagnostic characteristic of Social Darwinist literature: 'pale faces, stunted figures, debilitated forms, narrow chests and all the outward signs of low vital power'. Environmental causes were identified, such as the nervous strain and babel of noise of urban living; pollution of air and water; and, so far as children were concerned, over-pressure of school work, a campaign orchestrated in the 1880s by the medical profession. Moral decline was even more feared, and reflected the widely held stereotypes of the dissipation, self-indulgence and godlessness of the sub-cultures associated with the urban working classes.

The broader and more balanced elementary curriculum that gradually emerged from the 1870 Act (as compared with the Revised Code of the 1860s), once more revived complaints of over-pressurising the limited minds of the masses. A more appropriate inclusion for the mass of the people's children than academic subjects was seen to be military drill, designed, apart from improving the physique of urban children, 'to teach habits of obedience and discipline and...preserve law and order and diminish crime' (Hurt, 1977, p.171).

The shock of the evidence of stuntedness and physical degeneracy among recruits for the Boer War, and the reinforcement of this evidence by the Inter-departmental Committee on Physical Deterioration of 1904, was also more than enough to vindicate the extreme views of the eugenicists, who intensified the crusade to improve racial health. The First World War added a new anxiety through widespread outbreaks of venereal diseases among the troops, which spread through the population with alarming speed on their return to civilian life. Again the schools were offered the poisoned chalice of finding a solution. Domestic economy was thought by this stage to be unfitted to achieve the objective. This time it was science that provided the key. Biology took over from domestic economy and drill as the subject most likely to make a major contribution both to health and citizenship education. Information about sexually transmitted diseases could be offered naturally as part of the broader study of anatomy and physiology and the general coverage of infection. The language of moral as well as physical taint remained at the heart of health instruction, however. During the 1920s purity organisations like the British Social Hygiene Council argued for education for chastity in the schools. The Council sponsored a massive programme of public lectures, and VD horror films. The indoctrinatory shock tactics of this time were still apparent in the 1970s, approaches to sex education including, in Bristow's words, 'the nurturing of venerophobia, now augmented with films of giant magnified germs' (1977, p.230).

Geography, Social Education and the National Curriculum

The historical evidence suggests there is a compelling need to look more closely at contemporary sub-texts in cross-curricular areas, not least in the light of the politicisation of the curriculum by the government during the 1980s and early 1990s. The argument put forward in Chapter 9 was that issues-based approaches were necessarily cross-curricular. As the context of such approaches must also be political, it has to be accepted that a claim can be made that politicisation took place from both sides, both appealing to its own set of 'good causes', one an anti-socialist and the other, in its more radical forms, an anti-capitalist agenda. Subjects like geography and history find themselves in the centre of the discourse and, indeed, the polemic. Thus history 'from the bottom up' has, or certainly can have, a covert political agenda. The government's attempts to nationalise the history curriculum are more obviously overt (McKiernan, 1993). Where, therefore, lies geography's position, in the context of the National Curriculum, in relation to good practice in social education?

The Statutory Orders for Geography, and International Understanding

In its choice of personnel, and in the terms of reference laid down to the Geography Working Group (GWG), it was clear from the start that the Secretary of State's intention was that geography should fairly and squarely be defined as the study of places. The ambiguity of the term 'place' meant that some could interpret this as a realisation of a long-felt want, and others as a reactionary return to old-fashioned place knowledge. In fact it was neither. But by 1989 there had been significant and worrying back-tracking even on the hardly radical HMI *Geography 5–16* document. No explicit lead was given in the Secretary of State's brief to the GWG on the need to promote international understanding, though there were perhaps chinks of light in the references to the importance of covering 'the physical, economic, political and cultural relationships that link peoples living in different places throughout the world' and in leading pupils to understand the way in which people used the earth's resources 'in making their living and seeking to improve the quality of their lives' (DES, 1989b, p.86).

In general the *Interim Report* of the Geography Working Group received a bad press from the professionals. Psychologically, this may in part have reflected the fact that, unlike the history interim report, it was so well received by the Secretary of State, and was also tendentiously presented to the media by its Chairman as marking a return to fact-learning geography. More importantly, the Report was found guilty of the gaffes of dividing world geography into two parts – the developed and developing world – rather than stressing interdependence, of isolating the United Kingdom from Europe, and of specifying the Falkland Islands and various ex-colonial places as recommended areas of study.

In the Report's statement of aims, an appropriate amount of attention was devoted to issues of global interdependence, care of the earth and its peoples, and to promoting geographical enquiry. Stress was laid on links with certain cross-curricular themes, some of which, such as political education, were never again heard from any official body. Notwithstanding the deficiencies, hardly surprising in the light of the unseemly haste with which the Working Group had to produce its report, it offered a more wide-raging global agenda than anything to follow.

The broad aims of the interim report were reproduced almost verbatim in the *Final Report* (DES, June 1990). The main points of criticism about the place attainment targets were in general redressed. The subsequent National Curriculum Council's *Consultation Report* (November 1990), while containing further refinements and improvements, watered down the place element in the geography curriculum to one AT (2), defined as 'knowledge and understanding of places'. The fact that it did not treat the thematic targets in the same way reflected a body of opinion within the profession, well represented in the GWG, which felt it was important to retain a distinctive physical geography. To others, it represented a rather dubious reflection of priorities, reducing the proportion of content relevant to education for international understanding.

While the promotional rhetoric of geography as a subject uniquely qualified to promote international understanding was not satisfactorily implemented, at least the Consultation Document retained, especially in AT5, a viable issues-permeation and recognition of the relevance of an attitudinal dimension within such issues. Within a week of many receiving, at the beginning of 1991, the NCC Consultation Report (dated November 1990, but delayed at the printers), the Draft Orders (DES, 14 January 1991) appeared. No serious consultation had taken place, nor indeed could it have taken place on this time-scale. The third Secretary of State to emerge during the period of deliberations of the Working Group and the emendations of the NCC, Kenneth Clarke, had a more decisive view than his predecessors of what he wanted, and put right out of court any idea that there should be any appraisal of pupil's exploration of values and attitudes through the statements of attainment.

The specification of localities to be studied was further weakened, the European dimension almost disappearing from the primary phase on the grounds that the load at Key Stage 2 needed lightening. While the premise was correct, the choice of what was to go necessarily attenuated the input of material relevant to the cause of geography for international understanding. The Statutory Orders for Geography thus represented in this sense a sizeable victory for reaction. Notwithstanding potential improvements post-Dearing, outlined in Chapter 11, the net effect was, on the face of it, to reduce the possibilities of studying overseas localities still further.

Cross-curricular Guidance and International Understanding

The Statutory Orders for Geography were therefore increasingly heavily pruned of content relevant to education for international understanding as between Kenneth Baker's initial briefing to the GWG and Kenneth Clarke's Draft Orders, particularly in the all-important primary phase. As compared with the position in geography, the peripheralisation of international understanding in the cross-curricular National Curriculum Council Guidance documents was even greater. As a group these documents were disturbingly Anglo-centric in tone.

The absence of education for international understanding, or some equivalent, in 'Curriculum Guidance 3' on *The Whole Curriculum* (NCC,

1990b), self-evidently set the scene on what was to follow. Of all the contenders from which the cross-curricular dimensions and themes could have been selected, international understanding/global education and political education were the most obvious missing links.

Education for Economic and Industrial Understanding: Curriculum Guidance 4

Intellectually, the EIU document (NCC, 1990c) was perhaps the most impressive of the series. It clearly benefited from the conceptual spade-work undertaken by many projects promoting economic and industrial understanding. An appendix lists 36 projects and organisations engaged in this enterprise. Large amounts of money were expended by government, industry and commerce in this promotional endeavour. One of the organisations was the Department of Trade and Industry. Subsequent to *Curriculum Guidance 4*, a grant from the DTI allowed the National Curriculum Council to produce 11 further publications in this field. The 'partners in education' with the Employment Department, in producing an expensively bound publication for the Economic Awareness in Teacher Education (EATE) Project, were Boots, the Post Office and Grand Metropolitan. Some of the material is undoubtedly of potential educational benefit. But were there strings, however implicit, attached?

As already indicated, a preliminary concern relates to the linkage of economic with industrial understanding. While it might be argued that the development of economic understanding is a contribution to a liberal education, the addition of 'industrial' introduces a distinctly different orientation. It vocationalises and, intrinsically, politicises the agenda. The evidence from *Curriculum Guidance 4* would seem to reinforce this suspicion. As Carter (1991a) has pointed out, it is based squarely on a values position which echoes a free-market, western-consumerist view of the world (p.30). Nixon (1991) further reminds us of how the central messages from the document were interpreted by the press: 'Britain needs mini-Maggies in its schools'; 'Pupils will study business'; 'Entrepreneurs will start at five', and 'Capitalists in the classroom' (p.190).

A final piece of evidence lies in the examples that are offered to show how it is intended that the conceptual frameworks of the guidance, educationally valid in themselves, should be applied. It can be argued that these are diagnostic of the underlying thrust of the thinking, sending out most clearly the real messages to relatively unsophisticated generalist teachers. Thus in *Curriculum Guidance 4* the case studies are largely about running shops, superstores and mini-enterprises. There is some sound educational thinking in the frameworks, but little challenge to consumerism in the attitudes the document seeks to engender. The contrast between this narrowness and the expansive thinking of the Development Education Centre in Birmingham in linking geography, economic awareness, and international understanding is marked (Robinson, 1994).

Health Education: Curriculum Guidance 5

Health education is justified in *Curriculum Guidance 5* (NCC,1990d) as helping to protect children from illnesses and risks, to promote the development of healthy life-styles, in turn improving the quality of life and the environment. To succeed in what are no doubt laudable objectives requires a degree of intervention. There are different approaches to these goals and the balance between educational and instructional approaches needs to be clarified.

Again the more detailed advice in *Curriculum Guidance 5* on teaching approaches is diagnostic of the underlying thinking. It is true that children are not now expected to sign pledges asking for God's help in a lifetime of abstinence from the use of intoxicating liquor. Less blatant tactics are necessary. Yet the long-standing reputation of health education as an area adopting less than educationally appropriate scare-mongering tactics has still to be convincingly demonstrated as a thing of the past. We are socially and politically in a new era of purity rhetoric, and it is at least slightly disconcerting that in the framework of nine topics proposed for health education, represented in each of the four 'key stages' of the National Curriculum, the two pole positions are occupied by substance use and misuse, and sex education, both highly charged 'good causes', and suscept- ible to prescriptive and pathological slants in their presentation. In sensitive teaching hands, where the niceties of the distinctions between education and instruction are understood, this should not be a problem. But there is an element of prioritisation of those areas in which children are particularly prone to be proselytised. This is not to attack the basic premise that these must be crucial content in any sensible health education programme.

A second major concern is again how 'west-centric' the whole scheme is. The central thrust relates to health perils associated with the life-styles of wealthy citizens of western consumerist societies. Apart from the odd sparse reference, these is no feel that health education has an international dimension, and that most of the world's population is many decades from achieving even basic good health standards, an area of concern central to progressive geography syllabuses. Important global perspectives could have been noted under the topics 'Food and nutrition', or 'Environmental aspects of health education'. In the latter, knowing about the National Health Service is present, but not the World Health Organisation. If, in the spectrum from health education to health instruction, the Health Education guidance is not at the latter extreme, in the wrong hands it could well be unhealthily located on that side of the centre.

Environmental Education: Curriculum Guidance 7

By contrast, both in the Statutory Orders for geography, with the retained section on the environment, and in *Curriculum Guidance 7* (NCC, 1990e), on environmental education, the heady ambience of a new secular religion is all too apparent. Here issues of the ozone layer, acid rain, and the destruction of natural habitats, resource sustainability, and similarities and differences between peoples in the way they use their environments, have a genuinely global feel.

Again, however, much of the delivery at the global scale would appear to be left to geography. Most of the exemplification offered here is again Eurocentric, though there is one interesting exception in the suggestion of a Key Stage 2 approach to international links through a case study of twinning with a Nigerian school. What is also not made clear in the document, nor indeed anywhere else by the NCC, is a strategy for coping with the major overlaps between environmental education, as variously defined in this guidance document, and respectively in the science and geography orders.

There remains the paradox of a situation in which environmental geography on the one hand gains a prominent position in the geography Orders, confirmed by Dearing, and in which environmental education as a cross-curricular theme on the other is ignored (Goodall, 1994).

Citizenship Education: Curriculum Guidance 8

In the context of education for international understanding, the most disappointing document of all is *Curriculum Guidance 8* (NCC, 1990f) on citizenship education. Conjecturally, citizenship education was seen by the then DES as the theme with high-risk components, into which politically alien fellow-travellers would be most likely to permeate aspects of peace education, development education and global education. In a spectrum of definitions of citizenship, in a potential coverage from elementary school-type civics to a radical world citizenship approach, the document is manifestly more comfortable in the former neighbourhood. Machon (1991) notes that the theme of citizenship education does not here seem to be associated with an energetic and contested debate, and is more about 'subjects' than 'citizens' (p.128).

Many of the appropriate terms – a pluralist society, equal opportunities, global issues, and the need for international co-operation – are mentioned in the encompassing rhetoric. In terms of delivery, however, there can be no question that the document is rooted in a Victorian conception of citizenship, as evidenced by the practical suggestions offered. Thus 'progression' is illustrated by how to cover 'the citizen and the law'. The key concepts of rules and laws loom large. Geography is given some small responsibility for introducing a global element, as in studies of pluralism but, as Machon concludes, the prevailing tone is 'as British as a semi-detached house in the suburbs' (p.128). Perhaps the give-away section lies in the revival of the old-time call of the civics syllabus: to enlist the help of adults in work roles in the community by inviting them into the classroom to explain their work. While the idea is pedagogically sound, the choice of roles is hardly balanced: the organisations recommended are the St. John's Ambulance, the Red Cross, the Scouts and Guides, the Duke of Edinburgh Awards Scheme, Outward Bound, Operation Raleigh and, 'of the greatest importance', the Police Service, which 'can help in developing the ethos of a school' and 'support active, participative citizenship through enterprises such as Junior Crime Prevention Panels' (p.13). Whither Oxfam, VSO, UNESCO or, dare one mention, Amnesty International?

It is surely an evasion of responsibility in a country in membership of international organisations and subscribing to their mission statements, that the National Curriculum document on citizenship education's references to UNO, for example, are minimalised in the limited references to its role in 'work, employment, and national and international economics' (p.9), and its main features as an international organisation. There is also reference to covering major conventions on human rights, and comparing the situation in Britain on racial issues with that in other countries. Unlike aspects of internal concern, however, these matters of global import are peripheral to the main thrust.

So far as the National Curriculum cross-curricular guidance is concerned, therefore, and not least in citizenship education, we appear to have a case of allowing international understanding to wither on the vine.

Covering International Understanding?

While only the rank optimist could anticipate an easy future ride for education for international understanding in England and Wales (and more particularly in England), there are some grounds for hope. One lies in the dual economy of the National Curriculum, and the propensity this has for a fruitful blurring of the distinctions between subject and cross-curricular approaches. Much ideological blood, sweat and tears has been expended on fossilising and polarising these distinctions. There are different pathways to implementing the National Curriculum, which can be turned to advantage. The professional teacher will undoubtedly recognise the difference between the concept of a broad and balanced entitlement, statutorily laid down in the rhetoric, and the minimalism of some of the Statutory Orders.

In the light of the programmes laid down in the Statutory Orders, a major part of the delivery of key aspects of international understanding must inevitably be through geography. As has just been noted, there are many global left-overs for it to pick up. The problem remains that geography should not feel entitled to claim that it can fully cover these issues by itself. Most of the great international issues, as indicated in Chapter 9, are inter-disciplinary and cross-curricular in their scope.

Where then, can these broader issues be covered? A practical problem is the constraint of fitting all the cross-curricular dimensions and themes into the scarce time-slots remaining after the Statutory Orders have been covered. It is clear that the claims made on this time need to be carefully assessed, and opportunities taken to capitalise on the major overlaps between subjects and with cross-curricular elements. In the time not dedicated to the National Curriculum subjects, some extra might be offered to geography, for example, to lead a focused topic on an important aspect of international understanding that has both important geographical elements, but requires activity beyond the confines of the subject, including the co-operation of staff from other curriculum areas in the endeavour.

An alternative is to look at a more broadly co-ordinated topic under the aegis perhaps of personal and social education, which surely should introduce the global responsibilities of individuals. There remains a question as to whether this support for education in international understanding is compatible with arguments advanced earlier that many aspects of cross-curricular study are too readily open to inculcatory techniques, justified on the grounds that they are in support of a 'good cause'. Clearly in any humanistic view of schooling, education for international understanding does represent a good cause. A number of tests can be applied to avert the possibility of indoctrination. The teaching material should be checked as being balanced, qualitatively as well as quantitatively, and clear as to the hidden as well as the open agendas. Thus the use of material from *The New Internationalist*, for example, which appropriately is overt about the political slant of its presentation, can be used as a counterpoint to, for example, material from the glossy brochures of trans-national corporations, where the bias is sometimes more covert.

One of the saddest aspects of the National Curriculum, and most particularly in the Curriculum Guidance documents is, therefore, how they glow with a reconstituted nationalism. It was less so in official documents of the late 1970s and early 1980s, as we have seen in the *Geography from 5–16* (DES, 1986) document. The development of political literacy, particularly in the broad sense of international citizenship, is morally and culturally prescriptive, a point which is reinforced at international level, through global organisations to which the British government subscribes. Thus there are the Stockholm and Nairobi declarations on the environment; Council of Europe declarations on the teaching of human rights in schools; and UNESCO on the specific issue of education for international understanding, co-operation and peace, not to mention the declarations of various conventions on human rights. The general import of these declarations, if followed both in the letter and the spirit, is to preclude the temptation of governments to proselytise narrow sectional, political and national interests.

The need is for a broader agreement on an approach to education for international understanding which, while differentially implemented in accordance with national curriculum programmes, at the same time works within a global humanistic/human rights framework. The International Geographical Union has indeed made a start on this in its approval of an *International Charter on Geographical Education*:

'Education should be infused with the aims and purposes set forth in the charter of the United Nations, the Constitution of UNESCO, and the Universal Declaration of Human Rights, particularly Article 26, Paragraph 2, of the last-named, which states: "Education should be directed to the full development of human personality and to the strengthening of human rights and fundamental freedoms. It shall promote understanding, tolerance and friendship among all nations, racial or religious groups, and shall further the activities of the United Nations for the maintenance of peace".' (International Geographical Union, 1992)

If it is the current government's continuing intent not to give priority to the international education commitments it is already a signatory to, then teachers of international understanding must take the responsibility for seeing it does not wither on the vine, by percolating a global perspective into the joints of what fortunately are the highly permeable strata of the National Curriculum: whether in subject areas or in cross-curricular activities matters less. It is presumably a benefit of the freeing up of the curriculum by the post-Dearing slimming exercise, that teachers will hopefully be able discover a little more room to explore these wider horizons. But the commitment is more important than the time factor.

CHAPTER 11

Geography in the National Curriculum

This chapter deals with the events leading up to the selection of geography as a foundation subject in the National Curriculum, and with both the pre- and post-Dearing changes in the implementation of National Curriculum geography, following the 1988 Education Reform Act.

The late 1980s and early 1990s: Polarisation and Politicisation

The Influence of the Inspectorate

As earlier noted, the attack of the political right on the 'education industry' began in earnest in the early 1970s. Civil servants and politicians demanded an increasing influence in the school curriculum. When the Conservative government of Margaret Thatcher began its long term in office in 1979, the stage was set for what has been called 'the tightening of the ratchet'. The early stages of this process were exemplified in an increasing number of official papers produced by Her Majesty's Inspectorate. The 1978 *The Teaching of Ideas in Geography* (DES, 1978b) and comparable pamphlets in other subject areas appeared as harbingers of this new trend. The concurrent critique in *Primary Education in England* was also the work of HM Inspectorate and the Department of Education and Science (DES, 1978a). It was censorious about many aspects of primary schooling, not least the teaching of geography.

One important contribution to the debate was HMI's *Geography from 5 to 16: Curriculum Matters 7* (DES, 1986) which, while no doubt not going far enough to appeal to more radical voices in the field, none the less was progressive in urging attention to controversial issues and social and environmental concerns in geography. It stressed the need to provide programmes for all pupils which reflected good geography and good educational practice. It manifestly was influenced by the curriculum thinking of the 1960s and early 1970s. There was emphasis on principles of match and progression. It represented a more balanced and expansive view of curriculum renewal than later HMI statements, such as *The Teaching and Learning of History and Geography* (DES, 1989a), by which time the inspectorate's position was looking less than independent of government.

The Influence of the Geographical Association

For 100 years the Geographical Association, founded in 1892, has played a key role in support of geographical education. During the last two decades the Association has been very active in publishing guidance for teachers. Apart from its main journal *Geography* (which in its early years was called *The Geographical Teacher*), it also produces *Teaching Geography*, mainly for secondary school teachers and, more recently, *Primary Geographer*, targeted at generalist primary school teachers. In addition it has produced handbooks both for secondary (Boardman, 1986). and primary (Mills, 1988) phases. It has also offered a wide range of advice on, for example, field work, information technology, teaching slow learners, and on cross-curricular links, not least those between geography and technical and vocational education, and geography, schools and industry.

The Association would see as its main achievement in the last decade, however, its timely intervention to secure for the subject a place in the National Curriculum. In the Department of Education and Science's 1981 pamphlet on *The School Curriculum*, geography was not identified as a separate subject in the school timetable, though a geographical contribution to integrated studies was envisaged (Proctor, 1984). By 1985 the case for the subject was still not officially accepted. So the Geographical Association invited the then Secretary of State for Education to address a specially convened public meeting (Joseph, 1985). He asked geographers in turn to answer seven key questions as a basis for justifying the place of their subject in the curriculum. This the Association did in its *A Case for Geography* (Bailey and Binns, 1987). The publication arrived on the next Secretary of State's (Kenneth Baker) desk about the time he was deciding on the foundation subjects in the National Curriculum. He is said to have recognised that the geographical lobby had got its act together and accepted the case. Various accounts have been offered of the success of this campaign by Bailey (1989) and Walford (1992), and also of the work of the National Curriculum Geography Working Group, by Rawling (1992) and Morgan (1994).

The Pre-Dearing National Curriculum in Geography: a Return to Place

In the calculus of advantage and disadvantage, many geographers would claim that, for all its faults, the National Curriculum has strengthened the position of geography in schools.

Advantages

The National Curriculum has made explicit geography's:

- Position **as of right** in the National Curriculum for all pupils from 5 to 16, representing a major advance, at primary level in particular. The later amendment which removed this entitlement for the 14–16 age range is arguably a regression, though it does not necessarily alter the pre-National Curriculum situation, in which geography was optional.
- **Distinctiveness** in:
  - Returning **place** to the centre of the stage, focusing on detailed localities and other scales of place, and on spatial studies into which geographical themes and skills must be permeated. It thus binds physical and human aspects of the subject into places in an authentically geographical way.
  - Highlighting the importance of the skills of **graphicacy** and **field work**.
- Commitment to **enquiry-based learning**; though perhaps more strongly in the preambles than in the programmes of study.
- Capacity to function as a **bridging subject** in the curriculum, evidenced in the cross-connections made with humanities and science subjects.
- special contributions to **environmental awareness** and the development of **world knowledge**.

Disadvantages

There were, however, fundamental flaws in the National Curriculum Mark 1 for geography, widely recognised at the time, and later seized upon by the government, sensing that there had been an extravagant diversion from the basics. Conspiracy theories about the malign influence of the enthusiasts and experts of the education industry in undermining the government's reforms enlivened the pages of the right-wing press. This does not imply that some of the criticisms were justified, rather that the criticisms were being used inappropriately to support the case for a more intensive and instructional diet of the basics:

- The programme was **overloaded with content**, particularly at upper primary level (Key Stage 2), but also at lower secondary (Key Stage 3).
- The curriculum was **assessment-led**, which caused many to suspect that teachers would concentrate on meeting National Curriculum requirements by cramming in content rather than developing understandings and skills: that is, they would 'teach to the test'.

- This problem was made worse by the **absence of an explicit theory of learning** behind the orders for geography, which again could encourage teachers to build up programmes of study not by means of carefully thought out conceptual frameworks, but through an accumulation of content.
- The **statements of attainment were tied to content**. Thus arbitrary and unsustainable judgements were made about what should appear where and how: for example, that volcanic eruptions should be taught at Level 4, while the distribution of earthquakes and volcanoes in relation to tectonic plates was to be confined to Level 5, and so on. There seemed to have been little concept of a spiral curriculum (*see* Chapter 6), for example, in which it is axiomatic that each theme is revisited at different stages, at progressively more refined and complex levels of understanding.

In terms of our initial criteria, those who had sought a distinctive geographical contribution to the curriculum could in this sense be reasonably satisfied. The pre-Dearing Curriculum did demand distinctive geography. Whether or not it was associated with educational worthwhileness was likely to depend on whether teachers could avoid the negative influences of an assessment-led curriculum.

The Impact

Evidence of the impact of this first phase of the implementation of National Curriculum geography for Key Stages 1, 2 and 3 can be found in OFSTED reports on the implementation of the requirements of the 1988 Act for different subjects. In the second year of implementation (1992–3) for Key Stage 3 it was noted that:

- over 80% of lessons seen were at least satisfactory (as against 66% in 1991–2);
- 30% of lessons seen were good or very good;
- pupils more consistently recalled information well than they achieved satisfactory levels of understanding;
- geography was most often taught as a separate subject;
- assessment, recording and reporting were variable in quality and satisfactory in only just over 50% of schools;
- staffing was satisfactory or better in over 75% of schools and departmental management in over 66%;
- professional development/INSET opportunities were satisfactory in a little over 50% of schools;
- the resourcing of National Curriculum geography was satisfactory or better in 80% of departments;
- IT was too limited in its use in geography lessons.

The official impression was therefore that, for all its widely accepted flaws, the pre-Dearing National Curriculum had within 2 years of its implementation improved geography teaching at Key Stage 3.

Post-Dearing Developments

The defects had, however, by the time of the OFSTED reports, led to a general agreement that the National Curriculum in general, and subjects like geography in particular, were in need of radical overhaul. It was argued that each subject should identify the 'essential elements' that it felt should be retained after the Dearing cuts, and be prepared to lose the rest. The perceived essentials would provide the minimum entitlement for pupils at each stage. One point made by Dearing in the build-up to the review, however, was that in the necessary slimming down process the baby should not be thrown out with the bath-water. But what became worrying was that Dearing, SCAA, the Department for Education and ministers were under strong pressure from reactionary forces, some in so-called right-wing think tanks, some in government and, regrettably, some teachers, to pare down the curriculum to the absolute basics. Cynics referred to the 'five Rs' of government priority: reading, writing, arithmetic, right and wrong. What must be stressed is that while it is appropriate to regard reading, writing and arithmetic as basic skills, it is logically indefensible to infer from this that there are basic 'subjects'. The basic skills can and should be permeated into all subjects. For all the improvements suggested, the Dearing Report was a political document, picked up by SCAA, which was quite open that one of the influences in redrafting the proposals was its own concept of a 'minimalist' approach.

More considered arguments postulated a sensitive and sophisticated reduction, rather than wholesale excision of bleeding chunks. The key points identified by, for example, the Geographical Association (Carter, 1994), were that the changes should:

- lighten overload without sacrificing breadth and balance;
- simplify unnecessary complexities without reducing coherence;
- amend the technical language without losing the geographical distinctiveness;
- resolve anomalies and overlaps between curriculum areas;
- avoid overlaps between phases;
- proceed in an evolutionary way.

The 'essential elements' in maintaining the entitlement were obviously those offering a distinctively geographical input. Thus it was critical to preserve, in these terms:

- the place and space basis of geography from local to global scales;
- map and photograph interpretation;
- enquiry-based approaches, including field work;

- permeation of the identified themes into place studies, at different scales;
- stressing inter-relationships as between the themes;
- retaining the 'cube' framework of the Geography Working Group.

Clearly the Dearing Geography Working Group (DGWG) had to work to a brief, and there is no challenge to this in what follows. Among other points, the following were required in the geography review:

- a considerable reduction in the amount of content;
- a compulsory core of geography for each key stage, with optional material which can be taught at the discretion of the school;
- a cut in the number of attainment targets and a drastic reduction in the statements of attainment;
- retaining the 10-level scale;
- ensuring the rationale of the old Order underpins the new;
- meeting the proposed time allocations in ways less prescriptive than the old Order;
- encouraging a greater integration of the teaching of skills, places and themes;
- emphasising the interdependence between the themes of physical, human and environmental geography;
- retaining the requirement for locational knowledge, as well as knowledge and understanding of particular places and their relations to the wider world;
- ensuring the maintenance of the development of the skills of IT and field work in the current Order;
- establishing more clearly progression in geography across the key stages;
- removing major areas of overlap between Key Stages, but still provide access at each Key Stage to material in the programmes of study for those pupils working at different levels from most of their peers in their particular Key Stage.

In May 1994, the Dearing *Draft Proposals* appeared, and in November the more or less final draft of the new Statutory Orders. While there were changes following the consultation between these two events, the focus here will be on the new Statutory Orders. Key points will be highlighted, rather than a verbatim account being offered.

The New National Curriculum in Geography at Key Stage 3

Preamble

At Key Stage 3, pupils should be given opportunities to:

- investigate places and themes at various scales;

- describe and explain the changing characteristics of places;
- appreciate the interactions within and between physical and human processes and how they affect places and geographical patterns;
- understand the nature of environmental issues and analyse different approaches as to how they may be tackled;
- become aware of the global context within which places are set, how they are interdependent, and how they may be affected by processes operating at different scales.

Skills

The promotion of enquiry skills is emphasised, including:

- identifying geographical questions and issues and establishing an appropriate sequence of investigation;
- identifying the evidence required and collecting, recording and presenting it;
- analysing and evaluating the evidence;
- drawing conclusions and communicating findings;
- focusing on geographical questions such as:
  - What/where is this place?
  - What is it like?
  - How did it get like this?
  - How and why is it changing?
  - What are the implications?
- engaging in field work experience, as well as enquiry in the classroom.

In the new Orders, illustrative detail is offered as to how the statutory detail can be implemented, again under the somewhat anachronistic injunction, 'pupils should be taught to.' This includes:

- developing an extended geographical vocabulary;
- map making and interpretation, at a variety of scales, including Ordnance Survey 1:25,000 and 1:50,000 maps;
- use of globes and atlases;
- selection and use of visual and other secondary sources of evidence;
- use of appropriate graphical techniques to present evidence on maps and diagrams;
- a wide range of uses of IT, including CD-ROM.

Places

Here a choice is required of two countries, one to be taken from the nations of the developed world, and the other from the developing. For each of the two countries, investigation has to be made of:

- the physical and human features which give rise to the country's distinctive characteristics and regional variety;

- the characteristics of two regions of the country and their similarities and differences;
- the ways in which the country may be judged to be more or less developed;
- the setting of the country in its global context and how it is interdependent with other countries.

Thematic Studies

Nine themes are identified which:

- may be taught separately, in combination with other themes, or as part of the study of places;
- should be set within the context of actual places and have some topical significance;
- should, taken together, involve work at local, regional, national, international and global scales; and
- provide coverage of different parts of the world and different environments including the local area, the UK, the European Union, and parts of the world in various states of development.

The themes are:

- *Tectonic processes* – covering global distributions then either earthquakes or volcanoes.
- *Geomorphological processes*, and their effects on landscapes and people, either through river or coastal studies.
- *Weather and climate*, including the water cycle.
- *Ecosystems*, including one major world type of vegetation and how it is related to climate, soil and human activity.
- *Population*, including global distribution, reasons for population change and migration, and how population and resources are inter-related.
- *Settlement*, including studies of location, growth and the nature of individual settlements, functions, land-use and conflicts over land-use.
- *Economic activities*, including differences between primary, secondary and tertiary industries; the geographical distribution of one type of economic activity and how its distribution has changed.
- *Development*, including ways of identifying differences, the effects of differences on the quality of life of peoples, and interdependence.
- *Environmental issues*, including the evaluation of environments, conflicting demands on their use, planning for and managing environments and the potential unintended effects of such planning, sustainable development and conservation, and studies of either water supply or energy supply and associated environmental issues. SCAA also offers a matrix noting the scales at which different places and

themes are required by the Order:

- *Regional* – Country A, Country B and population theme.
- *National* – Country A, Country B and population theme.
- *International* – Country A, Country B, and development theme.
- *Global* – Tectonic processes and population themes.

The **Welsh Order** for geography is broadly similar to that for England. The most significant difference is that in the former, Wales is specified as a third country for integrated place study, in addition to a specified country from the European Union (a narrower choice than in the English Order), and one from an economically developing country. There are also some differences in the thematic studies, 'Development', for example, not being included as such in the Welsh Order, though one is introduced on 'Global Environment'.

Progression

Advice is offered by SCAA that pupils should be given opportunities to:

- Widen the range of geographical skills.
- Deal with them at increasing levels of complexity.
- Move from a limited to a broader range of locality studies at Key Stages 1 and 2 to larger scale studies (countries) at Key Stage 3.
- Widen their knowledge and deepen their understanding of a larger number of themes, at Key Stage 3 across the whole range of scales.
- Increase their awareness of the world beyond their locality at Key Stage 1, to the wider world as a context for places/topics studied at Key Stage 2, to the global context of places and how they are interdependent at Key Stage 3.

Level Descriptions

These replace the old Statements of Attainment and relate to the new single Attainment Target, **Geography**. They describe the types and range of performance which pupils working at a particular level character- istically should demonstrate. In deciding on a pupil's level of attainment at the end of a Key Stage, teachers should judge which Level Description **best fits** the pupil's performance. It is indicated that the great majority of pupils should be working at Levels 1–3 by the end of Key Stage 1, Levels 2–5 by the end of Key Stage 2 and Levels 3–7 by the end of Key Stage 3. Level 8 is available for very able pupils and, to help teachers to differentiate exceptional performance at Key Stage 3, a description above Level 8 is provided. The old Levels 9 and 10 are scrapped.

The Level Descriptions in the new Statutory Orders are rather fuller than those of the Dearing Draft Proposals. Let us sample the revised Descriptions at Levels 4 and 5, the overlap levels between Key Stages 2 and 3. It is to be noted that these are not itemised in check-list form, presumably to avoid the dangers of the ticking off boxes mentality in the assessment.

Level 4

Pupils show their knowledge, understanding and skills in relation to studies of a range of places and themes, at more than one scale. They begin to describe geographical patterns, and to appreciate the importance of location in understanding places. They recognise and describe physical and human processes. They begin to show understanding of how these processes can change the features of places, and that these changes affect the lives and activities of people living there. They describe how people can both improve and damage the environment. Pupils draw on their knowledge and understanding to suggest suitable geographical questions for study. They use a range of geographical skills drawn from the Key Stage 2 or Key Stage 3 Programme of Study, and evidence to investigate places and themes. They communicate their findings using appropriate vocabulary.

Level 5

Pupils show their knowledge, understanding and skills in relation to studies of a range of places and themes, at more than one scale. They describe and begin to offer explanations for geographical patterns and for a range of physical and human processes. They describe how these processes can lead to similarities and differences between places. Pupils describe ways in which places are linked through movements of goods and people. They offer explanations for ways in which human activities affect the environment and recognise that people attempt to manage and improve environments. Pupils identify relevant geographical questions. Drawing on their knowledge and understanding, they select and use appropriate skills (from the Key Stage 2 or Key Stage 3 Programme of Study) and evidence to help them investigate places and themes. They reach plausible conclusions and present their findings both graphically and in writing.

Critique

Advantages

1. So far as Key Stage 3 is concerned, it can be argued that some of the gains have been at the expense of Key Stage 2, where a more severe cutting exercise was demanded. The principle that the overload should be lightened has, however, been met more than adequately at Key

Stage 3, and a degree of breadth and balance has been maintained.

2. Similarly, much of the unnecessary complexity of the existing Statutory Orders has been reduced. The underlying balance between and the principle of the inter-relatedness of skills, places and themes, expressed in the 'cube' of the original GWG, has been retained.

3. There remains an emphasis on place, map work and other distinctive contributions of geography.

4. While not enough has been done to allay the suspicion that the politicians are still not at all keen to give any precedence to promoting a global outlook and international understanding, the situation is clearly better at Key Stage 3 than at Key Stages 1 and 2, and even more obviously so in geography than in other areas of the National Curriculum.

5. The reduction to one Attainment Target makes good sense from the point of view of legibility to parents. There is widespread agreement that the move away from a myriad of content-focused Statements of Attainment is a thoroughly good thing.

6. The confining of locality studies to Key Stages 1 and 2 at first seemed to imply that field work at Key Stage 3 was no longer vital. This fear has now been redressed in that field work is explicitly built into the skills requirement.

Disadvantages

1. The home region has disappeared from both Key Stages 2 and 3. This is contradictory in that it is accepted in preambles that there is a necessary regional scale between the local area and the national. Yet nothing statutory now appears between the concept of a detailed locality (Key Stages 1 and 2) and a country (Key Stage 3). Pupils are therefore expected to study two regions of a country abroad without necessarily having got to grips with regional concepts as applied to their home area, though in the non-statutory illustration at Key Stage 2 setting the home locality in its wider region is now included: at least a small mercy.

2. While it is perhaps understandable that the DGWG and SCAA were under unrelenting pressure to be seen to be rigorous in cutting, it seems a pity that the Key Stage 3 curriculum should statutorily be reduced to the study of two countries. It can be argued, for example, that there is a reasonable case for the study of a different country in each of the three years of Key Stage 3. As with localities, such studies can be enormously integrative, properly used, in permeating a wide range of themes in a specifically geographical way. Permeating the themes in this way by means of another country study **need not increase the**

load. The 'third country' could with advantage have been the United Kingdom, included for pedagogic and certainly not for nationalistic reasons, to allow pupils first to become acquainted with country studies through, for example, looking at contrasting regions and particular themes in the home country, say in Year 7. How to build in the themes into the study of places at the national level is a problem left for teachers to solve.

3. Whilst the Key Stage 3 thematic framework is better conceived than those at Key Stage 2, there remains a niggling element of prescription, albeit much less than in the old Orders. Thus:
   (a) For physical geography, it would have been preferable to ask that schools cover one aspect of each of the former strands of the atmosphere, biosphere, hydrosphere and lithosphere at least once at Key Stage 3. This would give more flexibility than the current demand that teachers select volcanoes or earthquakes, and rivers or coasts. Why should also they not be able to have the option of glacial processes, for example?
   (b) In what seems a particularly inappropriate piece of cutting, communications and transport has been removed from Key Stage 3. There is no longer the former logic that these were present at Key Stage 2. No doubt it might be argued that these topics will have to be covered in place work, but why should teachers be denied this important theme? A core and options framework would have avoided this problem.

It must be admitted, however, that most of these disadvantages are minor irritants than crucial issues of principle. They are noted here because in most cases teachers can do something about them, if they so wish. The nature of the new curriculum is open enough to allow personal sensible judgements to be accommodated through taking advantage of the ideas undelying the non- statutory illustration, as will be discussed in the next chapter.

4. The problems with the new **Level Descriptions** would, however, seem to be more fundamental. They are one of the ways in which the Dearing review has moved matters in the right direction: away from the previous myriad Statements of Attainment, which were fundamentally flawed, as we have seen, in being content-tied and not related to any theory of learning. The concept of 'best fit' is probably a good one, albeit very difficult to apply with validity and reliability. It is not enough, however, that the new arrangements represent the lesser of two evils. The criteria of good assessment practice need to be met. To illustrate one of the basic problems, Table 11.1 lists command words and qualifying adjectives at each of these levels.

Level 4	Level 5	Level 6
Begin to describe Begin to show understanding	Describe and begin to offer explanations	Describe and offer explanations
Suggest suitable geographical questions for study	Identify relevant geographical questions	Suggest appropriate sequences of investigation
Use a range of geographical skills	Select and use appropriate skills	Select and make effective use of a wide range of skills
	Reach plausible conclusions	Present conclusions that are consistent with the evidence
Use appropriate vocabulary to communicate findings	Present their findings both graphically and in writing	

Table 11.1

In fairness, it must be pointed out that these command words are detached from the content they are linked with. But as the content is very similar as between adjacent levels, it is the rather obscure differences between these command words that presumably are key elements in deciding the assessment:

(a) Taking all eight Level Descriptions, it is apparent that the amount of detail prescribed at each level is weighted, with an increasing number of statements as the ladder is climbed, an educationally suspect proposition.

(b) It is also educationally indefensible to suggest that different command words can reliably be attached to different levels. It must be stressed that children at Key Stage 1, in simple ways, can describe, explain, demonstrate a range of skills, draw conclusions, present findings graphically and in writing, and evaluate. These capacities might be demonstrated at very basic levels, but they are there. The current approach is therefore liable to reduce expectation, if taken at face value.

5. There are also difficulties of interpretation. How does one, for example, distinguish a plausible conclusion (Level 5) from one presenting conclusions that are consistent with the evidence (Level 6)?

6. There are manifest problems of moderation and validation to be faced. Different teachers have different standards and levels of expectation and different skills. What constitutes 'best fit' could become something of a lottery.

SCAA must clearly return to notions of what progression mean, well laid out in the HMI *Curriculum Matters Geography 5–16* publication, and work out how these elements, the basis of a spiral curriculum (*see* Chapter 6), can be linked with the content strands of the Programme of Study (though not of course with the point by point detail). Thus the descriptions pay little or no attention to the notion of concepts, differentiating concrete from abstract concepts, and the like.

Crucial to the exercise is working out of different levels of understanding, starting from a limited awareness of basic concrete concepts to a sophisticated understanding of these and more abstract concepts, and the increasing capacity to apply understandings in new situations, to solve problems, etc. The problem is that of working out the levels that come between, and the degree of generality at which these need to be expressed. The Orders recognise this and offer a useful starting point and an appropriate overall structure. But the detail of the structure has still to be rescued from its current obscure state. While 5-year stability is an entirely laudable objective, it would be a pity if it enshrined not only flawed but impractical practice.

The detail of this criticism should not obscure the positive point that almost all the moves are in the right direction, and that there is sufficient flexibility and freeing up of the arrangements in the new Statutory Orders to allow the rekindling of good practice in geography, pedagogy and social education. As will be argued in the next chapter, to do this the statutory elements, while crucial, must not necessarily be regarded as the the only relevant parts of the new Orders to consider.

CHAPTER 12

National Curriculum Planning at Key Stage 3

Pre-Dearing Planning in Geography

As indicated in the previous chapter, OFSTED, on the basis of evidence from the first 2 years of the implementation of the Statutory Orders for geography at Key Stage 3, have noted encouraging general progress. So far as the impact of the pre-Dearing National Curriculum in geography at Key Stage 3 from the classroom end is concerned, the Geographical Association recently (1993) produced a report by Fry and Schofield on *Teachers' Experiences of National Curriculum Geography in Year 7*, and the following section is based on this useful document.

Fry and Schofield brought together six case studies of curriculum planning for Year 7, and also a similar number of Key Stage plans covering years 7, 8 and 9. They pointed out that at this stage the plans were a first attempt, likely to be changed in the light of experience. Having said this, they argued that each outline plan on the face of it represented a genuine attempt to cover the full KS3 Programme of Study. One complication was that, because of the way the Orders were conceived, this meant in practice that the plans were designed to cover Statements of Attainment from Level 3 to Level 7, overlapping considerably with Key Stages 2 and 4.

Fry and Schofield found that many similarities could be identified in the case studies. The most common organising idea for units was a combination of themes, issues and regions. The similarity between the programmes evolved was also quite striking, with more of an emphasis on local work and the home region in Year 7, broadening out, for example, into European and global scales in Years 8 and 9.

The editors commented appropriately that the initial task of grouping statements together, ultimately to form the correct number of units to cover the time available in KS3, was a complex task, requiring the simultaneous juggling of a number of variables. The key variables again were places, themes and skills, that is the three dimensions of the planning cube formulated by the NCGWG. This was an encouraging finding in that a programme based on a consideration of isolated strands, and even more of Statements of Attainment, might have negated the benefits of pursuing the inter-connections between places, themes and skills.

The frequent presence of introductory units largely devoted to map skills in Year 7 reflected a more restricted approach. This links with the previous mention of the phenomenon of Year 7 children being switched off by an unrelieved diet of mapping work. There was indeed also criticism in the OFSTED report of over-emphasis on mapping work at the expense of other Attainment Targets, and of the failure to check whether such work had or had not previously been covered at Key Stage 2. Fry and Schofield suggested that in a period of uncertainty, teachers might well have turned to something as tangible and familiar as map work as a safe starting point.

The authors pay much attention to the problems of the Key Stage 2–3 interface, for there was a marked neglect of the associated issues in the case studies. It appeared that detailed joint planning was very much the exception rather than the rule. This they found not all that surprising, especially when the amount of work needed simply to get the KS3 courses themselves off the ground was considered. Until the time pupils arrived in Year 7 having completed KS2, Fry and Schofield considered the problem of detailed curriculum liaison in geography between phases as unlikely to be seen as a priority by departments. It has to be said, however, that the problems may well not only be organisational, but also attitudinal. There is, for example, more than a residual feeling at primary level that secondary teachers might prefer to see themselves as starting up pupils' geography programmes from scratch.

Fry and Schofield also make much of the problems of secondary school misconceptions about what is going on and what is being achieved in the primary phase. It is clearly inconvenient for secondary teachers to find their intakes arriving with very different levels of achievement in subjects like geography. Secondary geography teachers will all experience, later if not sooner, the prospect of pupils arriving in Year 7 having inconveniently completed KS2 at a standard well above Level 3. Hopefully, the more discrete post-Dearing Programmes of Study and the greater flexibility, with content no longer tied to Statements of Attainment, will improve the situation. But there will not be effective liaison unless cross-phase communications are well established. Whatever the changes, the post-Dearing curriculum at KS3 will have to provide meaningful work for pupils working at levels below 5. Indeed, under the new Orders Levels 1 and 2 will be included for KS3, to meet the entitlement of special needs children. At the same time it has to be ensured that work from KS2 is not simply repeated. Significantly, many new commercial publications for KS3 geography have appeared to take no account of the fact that much of the content at Levels 3 and 4, and even some at 5, would already have been covered at KS2.

Fry and Schofield sensibly suggest that it is difficult to see how the pre-Dearing National Curriculum geography could be covered without a high degree of individualised work being set. The same will surely be true in the post-Dearing period. The problem is not a new one. It has long been recognised that geography is not a subject in which ideas of linear progression may be applied. In their case studies, Fry and Schofield found that the most popular approach was the obvious one of emphasising the content of the lower levels in Year 7 and moving upwards in Years 8 and 9. This meant of course allowing pre-ordained official notions to dictate what represented appropriate content at different levels. This is another difficulty likely to be reduced by the post-Dearing changes, in which content is not tied to levels.

Even where a coherent overall structure was attained by simultaneously juggling with places, themes and skills, there remained the difficult task of building differentiation into individual units, as well as trying to ensure progression in ideas and skills as between succeeding units. Clearly a perennial problem, vital to the success of any spiral curriculum planning process, lies in the revisiting of concepts at progressively higher levels. This was manifestly not addressed by the old Statements of Attainment and, as suggested in Chapter 11, is not adequately catered for by the new Level Descriptions.

Fry and Schofield were pleased to find the use of key ideas or key questions to structure a particular unit had survived the introduction of the National Curriculum, helping to give units an enquiry focus, and produce active, investigational work, rather than passive learning. This is perhaps heartening evidence to support a fundamental contention of this book, namely that if teachers are truly professional, they will fairly readily find ways round official constraints. In fact, many of the Statements of Attainment, while intended for a different and more negative purpose, could creatively be rephrased as key ideas, or even as key questions. The problem of these Statements was of course that were not conceived of as a possible basis for progression, some arbitrarily being born and others dying out at particular levels, denying the possibility of revisiting.

New Prospects for Flexible Learning

As was argued in Chapter 6, individualised flexible learning strategies are likely to be conducive to, but are not in themselves guarantees of, good classroom practice. The following section is based on the assumptions that enquiry-based flexible learning strategies are educationally an important means of addressing, for example, fundamental educational problems such as match and progression, as well as reflecting criteria of good practice in geography and social education. It is argued that professional teachers will now be able see the way forward to marry the procedures of flexible, enquiry-based learning with meeting:

(a) the minimalist demands of the new Statutory Orders;
(b) the more stimulating prospects as identified in the preambles and the non-statutory illustration.

Pre-conditions for such an achievement are that:

- from the **geographical** point of view, the materials and the frameworks should be distinctive, coherent, as first-hand as possible, and presented in an attractive and authentically geographical way;
- from the **educational** point of view, the procedures and materials should demonstrate the necessary sophistication and differentiation of questioning skills that enable match and progression to be achieved and which, together with the geographical components, will stimulate interest through carefully structured, enquiry-based learning;
- from the **social** point of view, the procedures and materials should not only develop personal skills in a well-managed and constructive social atmosphere, but also promote a sensitive understanding of problem issues related to the social and natural environments, at a range of scales, from the local to the global.

Some Guidelines for Curriculum Planning at Key Stage 3

The above principles are of course easier to enunciate than to put into practice. They are not, however, merely theoretical, and represent inescapable routes to good practice. They are therefore built into the guidelines for curriculum planning at Key Stage 3 below, which include some material on overall syllabus planning and an element of more detailed advice within this. These must be viewed as guidance and as one possible approach among many. Guidelines are what they imply. They are in the nature of advice and suggestions, and not of authoritative pronouncements sent down from on high.

Reviewing Existing Activity and Resources

Presuming that good practice is being followed, it obviously makes sense to conserve what has been achieved following the introduction of the new Statutory Orders. As Fry and Schofield, as well as OFSTED have shown, much of the recent implementation at Key Stage 3 has been creative and constructive. New resources have been purchased. It may be that not all that has been implemented remains viable, but the new flexibility means that most will be. An audit of what has been achieved and what can still be used in the light of the new Orders is thus a sensible starting point.

Meeting the New National Curriculum Requirement

There are a number of elements which need to be considered:

- the Statutory content of skills, places and themes
- the Statutory preambles
- the Non-statutory illustrations.

In overall curriculum planning for Key Stage 3, it would be indefensible and illegal merely to plough crudely through the Orders, say by covering skills in Year 7 (and there has been a tendency to overweight mapping skills development at this stage), followed by places in Year 8 and themes in Year 9.

Skills, areas and themes need to be related back to the original GWG's cube principle, resuscitated by SCAA. In addition, as SCAA has already indicated, there is a need also to accommodate the different scales of geographical study. The sort of scheme suggested is based on the principle that the different scales should be introduced in each of years 7, 8 and 9.

Thus the sort of overall syllabus framework that would meet the new demands (as one of many possibilities) might be:

- *Year 7* – Skills and themes permeated into places at:
  - local and regional scales, e.g. weather, settlement location and growth;
  - European Union scale (including UK), e.g. economic activities;
  - International scale, e.g. ecosystems (a vegetation zone).
- *Year 8* – Skills and themes permeated into places at:
  - local and regional scales, e.g. river/coastal, managing environments, water supply or energy issue;
  - Country Study A (in the developed world);
  - International/global scale, e.g. population.
- *Year 9* – Skills and themes permeated into places at:
  - local and regional scales, e.g. settlement, land use, primary, secondary and tertiary activities;
  - Country Study B (in the developing world);
  - International/global scale (tectonic processes, hazards, development and interdependence).

An advantage of the new arrangements is that the ordering of the syllabus is flexible, no longer being tied to Statements of Attainment associated with particular Levels, which allows other criteria, as considered below, to be taken account of.

The Geographical Good Practice Requirement

In the first instance, in resuming an emphasis on place and, at Key Stage 3, on country studies, it is vital to avoid any return to the static, deterministic and cumulative coverage of countries and their regions that were the hallmarks of an earlier regional paradigm (*see* Chapter 3). In

Chapter 2, attention was drawn to the helpful trends at the frontiers of geography, which demanded a return to the binding logic of place as the central focus of the subject, and to evolving new and more dynamic regional approaches. This must involve not a compendious coverage of traditional variables such as relief and climate, vegetation and soils, economic activities, settlements, and so on, but follow an interactive model which includes, for example, inter-relationships, inter-dependence with other areas, local cultures, community values, potentialities and constraints on change, and the regional disparities and cultural values which condition external perceptions of the region.

A critical element in good geographical practice is the collection of appropriate resources. Place-based geography makes rigorous demands in this respect. In developing a country study, therefore, it is first of all necessary to have a good resource base for the country as a whole, and its placing in the wider continental and global setting. One important criterion might be to select a country which is studied as a foreign language in the school, which is likely to be France, Germany or Spain. All these are potentially good choices. But this criterion is not an essential one. Thus a teacher with very strong acquaintance with, and a collection of first-hand resources on, Italy, or Australia, or the Russian Federation, should be encouraged to make a choice on these grounds.

But let us assume the choice is Spain. For this and other European countries, an invaluable starting point (though not a finishing point) is the Collins Longman/Geographical Association *Resource Atlas*, which provides textual, cartographic, photographic, diagrammatic, and graphic material on each of: Location, Climate, Landforms, Population, Migration, Employment, Economic activities, Transport, Trade, Environment and Regions.

Let us assume that as part of the Spain topic, one of the two required regions chosen is **Catalonia**, in which a good deal of specific place detail is envisaged as illustration of chosen themes. Here the amount of material in the atlas is insufficient. Most, but not all, of the resources, should be distinctively geographical. It is important to remember also that we wish to cover more than the familiar geographical headings, and establish contact, as appropriate, with other curricular areas. The following would offer a more than adequate resource base. But to acquire them would need either visits to Catalonia or personal contacts there, such as a twinned school. If these cannot be organised, then it is suggested an alternative country should be investigated/chosen. The ability to provide a satisfactory resource base is, as already emphasised, a crucial criterion of choice. The materials specified below were in many cases acquired through a study visit and two personal contacts in Barcelona:

- *Maps*:
    - (i) Catalonia maps (note there is a splendid regional atlas of Catalonia, which covers for Catalonia similar features to the

Collins–Longman atlas of Spain).
 (ii) Larger scale maps of specific areas at 1:50000 or 1:25000 scale. While some larger scale maps may be available at agencies such as Edward Stanford (Longacre, London), it would be preferable to find a way of obtaining original Ordnance Survey-type materials from the Catalonian equivalent in Barcelona.
 (iii) Appropriate street maps of, for example, Barcelona.
 (iv) Appropriate historical maps.

- *Visual materials*:
 (i) Satellite photographs.
 (ii) Vertical and oblique aerial views.
 (iii) Selection of ground level photographs.
 (iv) Postcards and/or
 (v) Old prints and photographs.
 (vi) Video material.

- *Statistical information* – to provide a basis for:
 (i) Graphs of the trade of Barcelona, for example (based on materials obtainable from the Port of Barcelona Authority).
 (ii) Climatic graphs.
 (iii) Population charts.
 (iv) industrial development at primary, secondary and tertiary levels.
- *Timetables* – Air, rail and bus, to bring out contrasts between different purposes, distances and time scales.
- *Textual materials* – These could be first or second-hand and could include:
 (i) Original newspapers in the home language.
 (ii) Travel literature.
 (iii) Tourist guides.
 (iv) For the teacher, academic texts and articles.
 (v) Reports and brochures from commercial concerns.
 (vi) Novels about Spain and Catalonia.
 (vii) Dictionaries/phrase books – Spanish and Catalan.
 Links here with the modern languages department might be useful.

- *Memorabilia* – While some of these might not add a lot of geographical substance, they may well be helpful in establishing a sense of place and promoting cross-curricular connections:
 (i) Bus and rail tickets.
 (ii) Guides to museums, castles, etc.
 (iii) Wine bottle and other food/drink labels.
 (iv) Advertisements in the original language
 (v) Local postage stamps (obtainable from stamp dealers in this country).
 (vi) Local currency.

(vii) Local weather forecasts (in newspapers), and so on.

(viii) Local television and radio details (also in newspapers).

The effort involved in this wide collection will clearly pay off more if Spain is used as one of the key elements in dealing with skills and themes as well as the place requirement and also, perhaps, of linking with a modern languages course and, crucially, address to the cross-curricular European dimension. It is therefore re-emphasised that good flexible learning practices demand an equivalent priority being given to acquiring a range of resources that matches the quality that is regarded as the norm for the home area. That is what bringing the wider world into the classroom means.

The Educational Good Practice Requirement

This is placed fairly and squarely in the frameworks and principles previously advanced, namely of flexible learning strategies, of good questioning, of building in principles of progression and differentiation, and of capitalising on previous knowledge and understandings acquired, for example, at Key Stage 2. Good questioning has more than one element. We must think first in terms of distinctive geographical questions, and the new Orders appropriately offer a lead on this. But the lead might go further and be as precise as that presented in the HMI *Curriculum Matters Geography 5–16* publication (DES, 1986). In addition to the questions in the new Orders we might think of:

- Why is the place located where it is?
- Do few people live there or many?
- How has the place grown?
- What is it like to live there?
- In what ways are people's activities and ways of life influenced by the character of the place and its location?
- How are these activities distributed?
- What has produced this distribution pattern?
- What are its consequences?
- How have people made use of or modified the environment?
- Do many people visit the place and for what reasons?
- What important links does it have with other places?
- What is the nature of the routes linking these places?
- Which are the best routes to select, and why?
- In what ways is the place similar to,or different from, your own home area?
- What are the reasons for the similarities and differences?
- Is the place changing in character and, if so, why?
- What do you feel about the place? What do you find attractive or unattractive about it?

- Is the place changing in character and, if so, why?
- Do you think the changes taking place are an improvement or not?
- What are the views of the people who live there? (pp.14–16).

Above all, the questions should have content validity, that means they should assess worthwhile educational objectives and also what they are setting out to assess. Thus, to use Manson's check-list (*see* Chapter 6), there should be, in addition to the easy-to-set recall of information questions:

- insightful questions that promote thinking and problem-solving;
- valid and reliable questions that allow the teacher accurately to assess pupil achievement and diagnose problems;
- thoughtful and clarifying questions that promote self-awareness and self-evaluation;
- provocative questions that stimulate interest and group discussion.

Apart from the geographical dimension, the questioning must be appropriate in educational terms. In terms of **progression** the questions need to be related to notions of a **spiral curriculum**, in sequence reflecting, as we have already noted, increasing:

- breadth and scale of studies;
- depth and complexity of studies;
- abstraction of concepts covered;
- range of skills;
- complexity of issues.

The contribution of the new Orders creates some confusion, because while ideas of progression are certainly built in, the essence of progression is using the background force of the spiral curriculum, about revisiting and refining concepts, which is not.

In respect of **differentiation** (*see* Chapter 9) it is vital to:

- use a wide range of techniques of assessment, examples of which are given in that chapter; and
- think in terms of tiered or stepped assessments, based on differentiation by task, with carefully produced inclines of difficulty; or
- if following the strategy of differentiation by outcome, making available in addition to core tasks or activities, reinforcement and extension activities.

The Social Education Good Practice Requirement

As we have seen, the National Curriculum Planning process as a whole has left geography as the subject explicitly linked with current global issues, bearing a heavy responsibility therefore for the international understanding dimension of social education (including of course environmental education). A related matter is that work on global issues, even if geography-focused, is essentially cross-curricular in nature. Thus to meet the good practice requirement of social education, it has to be assumed that topics, even when geography focused, must take account of the cross-curricular themes, such as environmental and citizenship education, and dimensions, including the European dimension. One of the elements in curriculum planning for Key Stage 3 is also to build in the relevant aspects of cross-curricular skills (not least IT), dimensions and themes at appropriate points in the geography scheme of work.

It is especially important to avoid the perils of negative or otherwise distorted stereotyping. It should not be difficult to steer clear of the horror story stereotypes applied to Spain in older textbooks but there are other forms of stereotyping (*see* Chapter 8). A number of organisations during the 1980s drew up check-lists for appraising teaching materials, with the intention of guarding against the dissemination of racist and other stereotypes. The exercise was very necessary in that materials were still in frequent use as late as the 1970s and 1980s which did not take account of sensitivities related to class, age, gender and race.

(a) Check-lists – Check-lists are a useful means of auditing for stereotyping and bias, and the following reflects an adapted Development Education Centre check-list. It also helps to promote a dynamic geographical input as well. Note that this is a list directed at the global level of resolution, and requires some adaptation if being used as guidance for a country study, such as Spain. It would apply more directly to the List B countries of Latin America, Africa or Asia, as outlined in the new Orders in the Key Stage 3 places section.

- Ensure the material does not present physical factors in a deterministic way, and shows the physical environment as offering opportunities as well as constraints.
- Do not neglect the historical dimension which provides a vital context for understanding, especially in the context of post-colonial societies.
- Place the topic in a global context, as appropriate. In the case of Spain, this could be linked with the previous guideline.
- Discuss occupational structures and the types of work different people do.
- Explore the impact of post-modern technological change and time–space compression on different social and economic structures.
- Deal with population increase rates, densities, birth and death rates, and migration patterns in a comparative way and at different scales.
- Ensure a balance of different types of human settlements and their characteristics and functions.
- Indicate ethnic, gender, age and cultural diversity and the plural nature of many societies in the context of human rights declarations, and

ensure inclusive language is used.
- Demonstrate the achievements of minority cultures, e.g. Catalonia in the case of Spain as a whole.
- Deal with spatial variations in the levels of human welfare, measured on a variety of scales.
- Cover different measures of human welfare and their validity.
- Deal with the content and direction of international trade, and the impact of terms of trade on countries at different levels of development.
- Discuss the nature of aid relationships.
- Examine the economic, social and cultural effects of international tourism.
- Discuss the roles and economic and social impact of large trans-national corporations on countries of different levels of development and on regions within these countries.

Uncritical use of check-lists like the above can, however, produce slanted and doom-laden selections of material which themselves require auditing for balance.

The HMI *Curriculum Matters Geography 5–16* document of 1986 offers a useful corrective in this light to over-zealous interpretations of the nature and purposes of development education, suggesting a complementary check-list which can help to structure enquiry-based issues work in geography:

- What appears to be the nature of the issue and what are the geographical aspects?
- Which people and what places, locations or environments are involved?
- What is the relative importance of the geographical aspects of the problem?
- What views are held by individuals and groups about the problem and its possible solution, and how do these vary?
- What attitudes and values appear to underlie the views?
- What other information do you require to investigate the issue?
- How can this information be collected, collated and analysed?
- How good is this information as evidence?
- To what extent does the evidence support or contradict alternative views?
- What are the advantages and disadvantages of alternative solutions?
- Who would benefit and who would lose?
- What are your own feelings about the issue? Which proposal do you favour and why? What further information do you require to make a personal judgement?
- Have you changed your views as a result of your investigation? In what way(s)? (p.34)

(b) The European dimension – The European dimension is not secure in the new Orders at the level of places for Key Stage 3, in that a country study drawn from Europe is an option, but not compulsory. It is, however, statutory that so far as the thematic studies are concerned, topics must be covered in a range of contexts, including the European Union. As previously mentioned, it is a truism to suggest that in addressing themes associated with Europe as a cross-curricular dimension, links with other departments such as history and modern languages are useful.

In the European context, constructive official guidance has been published by the Department for Education (1992), on developing policy models for schools on the European dimension. The four broad categories of objectives proposed relate to helping young people to develop:

- knowledge and understanding of Europe, its peoples and its place in the world;
- positive but critical attitudes towards other peoples and cultures;
- respect for different ways of life, beliefs, opinions and ideas;
- enhanced language capability to facilitate communication and co-operation.

A curriculum audit is suggested, to review a school's existing provision. In the secondary situation, how can greater coherence be achieved through linking disparate offerings. Schools are recommended to identify this provision:

- within National Curriculum subjects;
- in subjects outside the National Curriculum;
- in cross-curricular elements, including personal and social education.

The European dimension, it is argued, should be part of a school's whole curriculum policy.

Different approaches are advised in implementing European dimension policies.

- permeating the whole curriculum, taught through the National Curriculum and other subjects;
- separately timetabled, as European Studies, with perhaps short modules on Europe;
- separately timetabled as part of a personal and social education course;
- short intensive programmes, involving a block of, say, one activity week on a European theme.

The document argues that the National Curriculum provides the opportunity to offer a co-ordinated approach on the European dimension, with some subjects – such as geography, history, art, music, and modern foreign languages – more centrally involved than others. The opportunities related to geography have been outlined earlier in this chapter. The cross-curricular themes of relevance to the European dimension identified by the Department for Education are the predictable ones: environmental education, economic and industrial understanding, careers education and guidance, and education for citizenship. Some schools are cited as having experimented successfully with teaching part of a geography or history course on a European country through the medium of the indigenous language. At least, materials involving use of the language should be available.

The DFE advice on learning experiences is formative in stressing the importance of relating work on the European dimension to the lives of young people, rather than offering second-hand materials didactically. Work on the European dimension is seen as more likely to be successful where:

- Direct experience of some kind is involved, such as:
  - a twinning link with a school in another country;
  - use of primary materials from abroad (e.g. products, currency, stamps, dance, music, etc.);
  - use of IT and electronic mail;
  - personal contacts;
  - use of satellite television;
  - visits/exchanges.
- Pupils are encouraged to participate in extra-curricular activities such as events linked to town and regional twinning.
- Employing, where possible, teaching assistants from a European country.

This guidance reiterates a number of points stressed previously in presenting ideas for a country study of Spain. The factor to be emphasised in conclusion, is that in fulfilling geographical, educational and social education good practice, the importance of teacher familiarity, of first-hand experience, is paramount. Much support is now available from agencies linked with Europe, to enable such resources to be collected and visits to be sponsored. These include:

The Central Bureau for Visits and Exchanges
Seymour Mews House
Seymour Mews
London
W1H 9PE

The UK Centre for European Education is also at the above address, as is the Council for Education in World Citizenship.

The European Movement
Europe House
8 Storey's Gate
London
SW1P 3AT

European Parliament Information Office
2 Queen Anne's Gate
London
SW1A 9AA.

Town Twinning Association
Local Government International Bureau
35 Great South Street
London
SW1P 3BT

A final caveat relates to the importance of maintaining balance as between enthusiasm for the European dimension and ensuring that, for example, not only is Catalonia seen in the context of Spain, but also Spain of Europe, and in turn Europe of the wider world. In the inter-war period, Thomas Pickles (1932b) certainly saw Britain as part of Europe and Europe as part of the wider world, but in these terms:

> '...the European has not been content merely to develop his own lands...the white man has with the irresistible drive of his energetic expansion discovered for himself, opened up, and then taken under control, all the continents of the world.' (p.1)

In its international role today, Europe has surely to pick up the legacy of this history. Also to be avoided is the traditional historical slant, in which European culture today is seen as the culmination of a privileged heritage derived from Graeco-Roman civilisation, enhanced by Christianity (Shennan and Lawrence, 1980, p.29). In the promotion of European awareness there is an uncomfortable prospect of presenting Europe in pan-nationalist terms, potentially as subversive of the promotion of international understanding as the nationalistic geographies of individual European nations in an earlier era. In this light, it is encouraging that the new Statutory Orders ask that the country study is set in a global context and also lay some emphasis on the issue of inter-dependence.

CHAPTER 13

W(h)ither Geography 14–16?

The National Curriculum at Key Stage 4

With the benefit of hindsight, it can be argued that the recommendations of the Task Group on Assessment and Testing (TGAT)(1987) set National Curriculum and GCSE assessment frameworks on a collision course. It suggested that the four top levels envisaged for the National Curriculum, 7 to 10, should equate with GCSE grades A to F. TGAT was anxious to avoid a situation of two over-lapping systems, and recommended in the first instance there should be only one reference point: namely that the boundary between levels 6 and 7 should correspond with the F/G boundary for the GCSE. An idea for later development was that the higher levels should be retained for those 14–16 year olds who did not wish to take a GCSE in, say, geography but who, under the proposals of that time, would have been required to continue with all the National Curriculum subjects until the age of 16.

In addressing the Key Stage 4 curriculum, the Dearing review was aware of the pressure of GCSE boards to conclude the National Curriculum levels system at the age of 14. The government response to the Interim Report was to agree that the GCSE would remain the vehicle for assessing 16 year olds in National Curriculum subjects and that, subject to the outcome of negotiation on the ten levels, GCSE results would continue to be graded on the old A–G scale, with an A* grade for the ablest students.

In the period between the TGAT Report and the Dearing Review the problem of the overcrowding of the curriculum had, as we have seen, been widely recognised. This was true not least at Key Stage 4. At this level the National Curriculum Council, among others, claimed that there was a compelling argument to offer a wider range of subjects in the GCSE than the core and foundation subjects of the National Curriculum. As early as the Interim GWG Report of October 1989, the then Secretary of State, John Macgregor, was inviting the group to undertake the 'necessary task' of developing a Key Stage 4 course for those who would not take the full GCSE course in geography.

The Final GWG Report recommended that Key Stage 4 should cover Levels 4 to 10. It specified how the then seven attainment targets could be approached by 16 year olds taking respectively the long and the short course in geography. For the latter group, while skills would be compulsory, there would be an optional element in both place and theme attainment targets. The Report suggested that the stronger focus on the study of particular places in the National Curriculum would require changes in the National Criteria for geography, to their benefit. The Group proposed a substantial course work element should be retained, with a weighting at the lower end of the current 20–40% range. Concern was expressed at the over-burden the course work component was said to inflict on individual pupils, and the opportunity for plagiarism and excessive parental involvement. It also argued strongly that schools should not see geography and history as the only possible combination of shortened courses, and that assessment of the short course should be as rigorous as that of the long course.

In November 1990, NCC recommended that all Key Stage 4 pupils should study geography either an individual subject or as part of a combined course, agreeing that the combination did not need to be geography and history but might, for example, be geography and economics. In May 1991 a separate Draft Order was produced detailing the content of the short course in geography at Key Stage 4. It was intended to apply from August 1994.

Following the appearance of the Statutory Orders and the ring-binders, in the case of geography in March 1991, the next phase was one of criticisms about over-loading and the assessment procedures as expressed in the over-numerous Statements of Attainment. Among other things, the situation at Key Stage 4 came under scrutiny. To cut a long story short, in October 1993 the government announced that the implementation of the National Curriculum at Key Stage 4 would be delayed pending the Dearing Review. By then, and confirmed in early 1994 by the Secretary of State of that time, John Patten, it had decided to lift permanently the requirement for Key Stage 4 pupils to study geography and history, which he argued reflected the Dearing review's recommendation that it would be desirable to provide more scope within the 14 to 16 curriculum for schools to offer a range of academic and vocational subjects corresponding to the differing aptitudes of pupils. An important issue was whether work in the GCSE in a subject such as geography could be transferred as credits towards GNVQ accreditation, with its modules in such geographically oriented topics as leisure and tourism, manufacturing, and the built environment (Butt, 1994).

In its response to the Dearing draft proposals, the Geographical Association made clear its concern about the removal of the National

Curriculum geography entitlement at Key Stage 4, arguing that it undermined the concepts of breadth and balance which the National Curriculum purported to offer, restricted the achievement of a full and rounded curriculum experience, as well as potentially affecting the health of the subject in secondary schools.

Beyond Key Stage 4: The New GCSE

By this stage, GCSE boards had been planning new Key Stage 4 GCSE syllabuses that met the requirements of the National Curriculum (Orrell, 1994). One of these was that there should be instituted tiering arrangements, each covering at least three but not more than four levels. Thus there might be a paper in a subject aimed at levels 4–6, one at 5–7 and one at 7–10. This particular variant would create problems of overlap and consequent complications for examiners and candidates. An alternative possibility, with the tiered papers aimed at levels 4–6, 6–8 and 8–10, avoided this overlap, put candidates under less pressure and allowed examiners' to set more interesting questions. Its disadvantage was the fraught choice for teachers in deciding precisely for which tier the candidate should be entered. Under this system, the differentiation was by task. It has been argued that many secondary teachers were moving towards a situation in which their Key Stage 3 implementation was leading naturally into the new National Curriculum GCSE syllabuses, and were more than a little displeased by the removal of the original arrangement by which there would be a compulsory Key Stage 4 geography examination.

SEAC Criteria

Arguably, there has long been a National Curriculum in place in the 14–16 phase, in that the GCSE boards, whilst retaining autonomy in their subject syllabuses, have had to work to National Criteria. In January 1993 SEAC issued Key Stage 4 examinations' criteria, in the context of a situation in which it was still envisaged there would be short and long course alternatives for geography. The structure was similar to that for previous statements of National Criteria to which the GCSE boards had for some time been complying.

Aims

SEAC demanded that a syllabus should:

(a) give pupils opportunities to acquire knowledge and understanding of a range of places, environments, spatial patterns and distributions;
(b) promote an understanding of the processes – physical and human – which affect their development;

(c) help pupils to appreciate that the study of geography is dynamic, not only because geographical features and patterns change, but also because new ideas and methods lead to new interpretations;

(d) enable pupils to acquire and apply appropriately the skills and techniques, especially in map work and field work, to conduct geographical enquiries;

(e) develop pupils' awareness of the ways in which people interact with their environments and appreciate the opportunities, challenges and constraints that face people in different places; and

(f) help pupils to act in a correspondingly informed and responsible way.

Assessment Objectives

These were closely tied to the geography Orders of the National Curriculum and, of course, became anachronistic in consequence of the Dearing recommendations, requiring the newly constituted School Curriculum and Assessment Authority (SCAA) to effect changes. One important detail was that in the schemes of assessment, course work had to cover not less than 20% and not more than 25% of the weighting, an attenuation of previous ranges in GCSE syllabuses. Each scheme of assessment had to involve an external examination, and weightings were given for coverage of skills, knowledge recall, and understanding, each to have not less than 20%, and not more than 40%, across the syllabus as a whole.

SCAA Criteria, 1994

SCAA issued revised criteria in response to the Dearing recommendations early in 1994, as a basis for consultation. They covered the subject-specific essentials, and also embodied SCAA's general requirements, including for GCSE examinations. The criteria were intended to build on the work achieved at Key Stages 1 to 3 in the National Curriculum, and also provide a foundation for more advanced work.

Aims

These were similar to those of the earlier SEAC document.

Assessment Objectives

A syllabus must:

(a) require candidates to demonstrate their ability to recall specific facts and demonstrate knowledge, including locational knowledge, related to the syllabus content across the range of local, regional, national, international and global scales;

(b) show an understanding of the geographical ideas, concepts and generalisations specified in the syllabus, and an ability to apply this in a variety of physical and human contexts;

(c) show an understanding of the range of physical and human processes which contribute to the development of spatial patterns and the geographical characteristics of particular environments;

(d) describe and offer explanations for the interrelationships between people's activities and their environments;

(e) show an understanding of the significance and effects of attitudes and values of groups and individuals involved in issues with a geographical dimension, and in decision making about the use and management of environments;

(f) appreciate the limitations of geographical evidence and the tentative and incomplete nature of some explanations;

(g) select and use a variety of techniques appropriate to geographical enquiry, including investigation in the field and, in particular, to observe, collect, record, classify, represent, analyse and interpret data;

(h) use a range of source materials, including maps at a variety of scales, photographs and simple statistical data; draw maps and diagrams; select, use and communicate information and conclusions effectively.

Syllabus Content

Syllabuses should:

(a) show a balance between physical, human and environmental aspects of the subject;

(b) give candidates opportunities to study a range of places in different parts of the world and in different types of environment, and to consider their interrelationships;

(c) give candidates opportunities to study a range of themes, at different scales (local, regional, national, international, global), in different parts of the world and in different types of environment: contexts for thematic studies must include the United Kingdom, Wales (for candidates in centres in Wales), the European Union and countries at contrasting levels of development;

(d) require studies of the interrelationships and interaction between people and their environments across a range or scales;

(e) require a first-hand study of a small area;

(f) require the study of the geographical aspects of important social, economic, political and environmental issues.

Systems of Assessment and Assessment Techniques

These reflected the recommendations of the earlier SEAC statement of National Criteria.

Current GCSE Syllabuses

It is useful to view current GCSE syllabuses, all following earlier National Criteria, in the light of the new recommendations, and in the context of best practice. Will the GCSE to come gain or lose from the new National Curriculum criteria, or is there likely to be little difference? The history of 16+ examinations since the early 1960s has been one of improvement.

Improvements in Examinations at 16+

Four improvements in particular can be identified.

The original CSE boards had to grapple with reactionary comment from some geography teachers who contended that to impose universal field-work for external examination purposes was impractical. To their credit, the CSE boards regarded such attitudes as tantamount to excluding children from an activity essential to progressive geography teaching. Schools found they could insist on field work, and the notion of field and associated course work as a result was enshrined in 16+ examinations.

A second major improvement was effected by the Schools Council Projects of Avery Hill and Bristol. As we saw in Chapter 3, these recognised at an early stage that however sensitive the projects were to the need to avoid top-down imposition, any 14–16 project would only become widely accepted if it tied itself into the external examinations system. In consequence, the progressive ideas of these projects were more widely diffused. The examinations improved enormously as a result, and specific syllabuses related to these projects were set in train, and continue to be popular.

The third significant improvement came from the merging of GCE and CSE examinations in the GCSE, again the subject of reactionary resistance. But it was clear that having separate GCE and CSE systems in a comprehensive secondary school situation set up debilitating and intrinsically unnecessary choices at 14+. Problems of how to cope with the high range of ability represented through one examination ensued. There was much concern that standards would be eroded. Imaginative and effective responses have in the event been made.

A fourth improvement resulted from the merging of examination boards. This began with the joining of GCE and CSE boards, and later from amalgamations of GCSE boards. Currently in England there are now four groups:

- Midland Examining Group (MEG)
- Northern Examinations and Assessment Board (NEAB)
- Southern Examining Group (SEG)
- University of London Examinations and Assessment Council (ULEAC)

Differential Practice of the Examining Boards

A present cause for concern is the fact that standards do vary, thus lowering official and public credibility. A comparative study of GCSE examinations in geography for 1993 organised by NEAB on behalf of all the boards (including also the Welsh and Northern Ireland boards) showed, in relation to a number of factors, NEAB to be somewhat less demanding, the Northern Ireland somewhat more, with the Welsh Board about average.

There were also differences in the assessment practices. Three Groups, NEAB, SEG and ULEAC, had common papers which differentiated by use of inclines of difficulty within questions and papers, and/or by the use of differentiation by outcome. The remaining three Groups offered different routes through the examination dependent on the target grade of the candidate.

MEG offered three distinct routes through the examination. Route A consisted of Paper 1 and course work and was intended for candidates in the ability range of grades D to G. Route B comprised Paper 2 and course work and was designed for candidates in the ability range C to E. Route C was intended for candidates in the ability range A to D and consisted of Paper 3 and course work. There was no common paper. Each route was weighted equally with 75% for the written component and 25% for the course work. The duration of each paper was two and a quarter hours.

Amounts of course work demanded varied. Two of the Groups, the Northern Ireland and the Welsh, required one piece of such work to be produced. Both Groups suggested a maximum length for the geographical enquiry of between 1500 and 2000 words. Three Groups (MEG, SEG and NEAB) provided the opportunity to produce one extended piece of course work or several shorter pieces of work MEG and SEG limited this to 2 or 3 pieces of work, whilst NEAB had no suggested limit to the number of separate pieces of work. SEG and NEAB offered a maximum of 2000 words. MEG stated no maximum length for the course work. ULEAC alone required two units of course work, one relating to the compulsory core of the syllabus and one to the option chosen for study. No suggested maximum length of the course work was given. In all the groups at least part of the course work was to be based upon field work. The weighting of the course work component(s) ranged from 20% to 30%.

Although all the syllabuses met the requirements of the National Criteria for Geography, there were variations in the interpretations of the content. Five of the syllabuses were non-location specific, allowing teachers freedom of choice of which locations to use for each topic, at a variety of geographical scales. NEAB, however, was different in this respect in that it was location-specific within its regional studies of the UK, the European Community and the Wider World. Nevertheless, the subject content studied at each of the specified locations was based on six topics which were similar to those offered by the other Groups. The six topics were: population and settlement; agricultural systems; energy and natural resources; secondary industry; tertiary industry; and transport and trade.

The content of the MEG syllabuses was centred on three themes: people and the physical environment; people and places to live; people and their needs. The subject content of the SEG was assessed in two main elements, skills and the units for study. The candidates were assessed by a skills paper (Paper 1) and in a paper covering the four units of study (Paper 2). The units of study for Paper 2 were: population and settlement; resources and economic activities; physical environment and human activities; development. ULEAC had a compulsory core of three themes: agriculture and industry; population and settlement; and landscapes and water. In addition it offered five optional areas: transport; energy; leisure; human welfare; the atmosphere and people.

The position was complicated by Groups providing alternative syllabuses. Thus the MEG Syllabus 1 emphasised the UK, and included the EC in its course work. Syllabus 2, conducted in association with the Welsh Board, constituted the Avery Hill Project syllabus, which emphasised issues-based and people-environment approaches. Syllabus 3, linked with the Bristol Project, focused on content associated with enquiry and decision-making activities, as part of a modular structure.

The uncertainty created by the switch of policy at KS4 created difficulties for those Groups planning new syllabuses to cover National Curriculum requirements at that level, and building on KS3 programmes. As Orrell and Wilson (1994, pp. 90–1) have indicated, many teachers were disappointed when the possibility of directly building on their KS3 work disappeared, or at least was shifted into the future. Nevertheless, MEG's new Bristol Project modular syllabus was made available for 1994–6 alongside the old one. This has anticipated that key features of the now defunct Key Stage 4 syllabus will be retained in whatever replaces it. The syllabus is distinctive in a 25% package on the home region, covering settlement and landforms; an ingenious externally marked decision-making exercise using pre-released resources based on environmental geography, worth 20% of the marks; and finally six themes assessed by enquiry-based problem-solving questions.

Grade Descriptions

Descriptions for grades F and C are provided in the National Criteria for Geography. These give a general indication of the standards of achievement likely to have been shown by the candidates awarded particular grades (Table 13.1).

	Grade F	Grade C
Ability	For Grade F, the student is likely to have shown the ability to	For Grade C, the student is likely to have shown the ability to
in relation to knowledge	recall basic information relating to the syllabus content and demonstrate an elementary level of locational knowledge.	recall a wide range of information in sections of the syllabus and, in doing so, will reveal a basic level of locational knowledge.
in relation to understanding	demonstrate a comprehension of simple geographical ideas; describe simple geographical relationships.	show not only a comprehension of important geographical ideas, concepts, generalisations and processes as specified in the syllabus. but will also demonstrate this comprehension, where required, in a range of situations, social, economic, political and environmental. The candidate will also be able to describe and account for inter-relationships between people and their environment.
in relation to skills	be able to observe, record and attempt to classify geographical data; to use a range of source materials, including maps; to draw simple sketch maps and construct diagrams such as a bar graph, to communicate information by brief statements.	be able to select data from a variety of sources, primary and/or secondary, to plan logically a simple geographical enquiry from observations to ultimate conclusion. Techniques used may include map interpretation at different scales, photographic analysis and a range of graphical and numerical information such as flow-line diagrams or simple census extracts.
in relation to values	recognise, at an elementary level, the claims of differing systems of values which influence economic, political and social issues.	show an increased comprehension of judgements made on economic, political, environmental and social issues.

Table 13.1

Alone amongst the Groups, the SEG also provides grade descriptions of the standards of achievement likely to have been demonstrated at grade A (Table 13.2).

For Grade A, the candidate is likely to have shown the ability to:

In relation to knowledge	recall detailed facts from across the syllabus, using accurate geographical terminology, and to be able to name good illustrative examples including relevant locations at the appropriate scales;
in relation to understanding	be able to support his/her comprehension of the essential elements of the syllabus by explaining causal interrelated factors, extrapolating likely developments, applying principles learnt in one situation to another, analysing conflicts, and by providing a synthesis;
in relation to skills	use appropriate analytical techniques, interpret maps and photographs to show an accurate "feel" for the landscape, draw concise and accurate sketch maps and diagrams and demonstrate the ability to select, classify, and interpret information in order to arrive at well-reasoned conclusions;
in relation to values	show reasoned understanding of differing viewpoints and evaluate those that could be accepted as both justified and feasible before coming to a conclusion.

Table 13.2

These grade descriptions clearly form the templates for the Level Descriptions planned for Key Stages 1 to 3 of the National Curriculum for ages 5 to 14. One advantage of the above descriptors is that they relate to a range of four specified types of ability. The technique of applying qualifying adjectives is widespread, as illustrated in the different demands at levels A, C and F on content. This is as far as the boards go, however, and what is expected between A and C and C and F is not specified, suggesting this is a difficult area. Clearly differentiation is still being effected at the detailed level of marking schemes.

In general, the new SCAA National Curriculum criteria, perhaps predictably, are shifting the balance away from:

- issues towards places
- higher levels of understanding and assessment of attitudes to assessment of locational knowledge
- course work to terminal examination, as evidenced by the statutory weightings.

The trend is not, however, dramatic, and there is still enough element of flexibility to allow the professional teacher to select materials and approaches that reflect good practice in geography, pedagogy, and in social education. It is also perhaps over-pessimistic to assume that the removal of geography and history as compulsory subjects in the 14–16 phase will have catastrophic effects. If geography teachers continue to show their capacity to motivate students, in what has the great benefit of being a socially relevant and potentially engrossing subject, then high rates of take-up at GCSE level are, as in the past, likely. Clearly in the new situation geography has to compete with the presence of imposed core subjects, but may be encouraged in terms of the popularity stakes by now being part of a recognised portfolio of examination subjects of which students have had statutorily laid down experience in the 5–14 age range.

At the vocational end of the scale, there is also the competition from the GNVQ. GNVQs have attractive pedagogic features in their demands for portfolios of work, the flexibility of modular options, the inclusion of self- and peer assessment, frequent interim appraisals allowing feedback, and other reflections of the good practice promoted by the profiling/records of achievement movement, leading to a cumulative assessment record (*see* Chapter 7). An elaborate verification procedure is involved. Their modules also reflect 'real-world' topics and concerns. On the negative side, there is arguably in the assessment the tendency towards too rosy a view of performance and competency-based criteria and the box-ticking mentality this can promote. There are also problems of internal quality control that have received much attention from OFSTED, forcibly presented by the press.

While geographers have been at the forefront of cross-curricular innovations related, for example, to economic and industrial under-standing and TVEI, the nature of the GNVQ modules implies not only cross-curricular but also such approaches in association with what could be a narrow vocationalism, carrying real dangers of loss of distinctiveness and of intellectual purpose (Butt, 1994).

CONCLUSION

CHAPTER 14

Approaching the Millennium

Decades of Disillusion

It is difficult for those who have for a working lifetime been associated with the world of education, and who can remember the relative liberality of the 1960s period, to look back over the last two decades in a positive spirit. For all the problems present in those earlier times, there was an overriding sense that education was a good thing, and that progress was the normal direction of change. The political context of education was radically altered, as we have seen, by increasing intervention from government from the 1970s. It was not so much the principle of intervention that was disillusioning, but the generally destructive spirit in which it was and has since been implemented, starting with the attacks of the far-right of the late 1960s and 1970s in the so-called *Black Papers*. For a time official intrusion was more cautious. This caution was gradually then more rapidly thrown to the winds in the 1980s. Thatcherist ideology brought in its train instability, confrontation, buck-passing and a strident moralism, not least in the negative stereotyping of the educational professions by right-wing politicians and press. In the approach to the millennium, educational progress will undoubtedly be checked, not least by a flight of teachers from the profession, if we do not return to some stability (psychological as well as curricular), common-sense judgements, and decent human relationships. It would be prudent in any case to accept Rawling's view that professional skills in future must encompass a higher degree of political sophistication (1993, p.111).

The Decade to Come

The National Curriculum, Teacher Professionalism and Teacher Education

At the same time, exciting developments have taken place during the last decade. Despite the deep-seated reservations previously expressed about its deficiencies, the National Curriculum is a momentous development, which in certain respects bodes well for the future, so long as a more constructive, enlightened and co-operative relationship is achieved between what should be partners rather than opponents in the educational enterprise. As Naish has argued, in relation to a larger project on the impact of centralisation of curricula in 22 countries, an agreed framework as part of such a centralised curriculum can 'encourage a common body of knowledge, ideas and skills without stultifying teacher creativity' (1992, p.49).

A continuing thread in educational thought over the centuries has been an agreement that the prime factor in achieving good practice is teacher quality. Those defeatist about the impact of the National Curriculum must presumably have an image of a teaching force made up of rule-following, narrow craft operators, who will be minimal in response and submissively allow the assessment provisions to drive the curriculum. The alternative view is that there are more than sufficient autonomous professionals in the geography teaching force not to allow this to happen. This cannot be a confident judgement, for the existing political disenchantment has a negative psychological effect, even on the most professional of teachers. In geography in particular, however, it would be my conjecture that the culture imprinted on specialist teachers over the last two decades, for all the problems, has been in the direction of progressive, enquiry-based learning. If this is the case, many if not the majority of teachers will subvert any return to a narrow, didactic, 'back to basics' tradition of, for example, inculcating locational-type information. There was obviously a risk that the over-weighting of the pre-Dearing Curriculum would inhibit broader approaches. The great advantage of Dearing, as we have seen, is that in geography it frees up the assessment system, and offers a greater degree of flexibility in its frameworks (Daugherty and Lambert, 1994).

Another threat to teacher quality, however, lies in the attitude of government towards teacher education. Its current policy is to shift the responsibility from institutions of higher education to the schools. Many institutions have long promoted and cherished partnerships with schools. School-based work in the University of Liverpool Department of Education, for example, was taking place before my own appointment in 1970. There would be little or no objection to making such contacts a more equal partnership than perhaps they have been in the past. This is not enough for government, however, which seeks to marginalise institutions of higher education as little more than bureaucratic co-ordinators and auditors of the new school-centred model.

There is already circumstantial evidence to suggest that this is a regressive development. However well schools undertake student education, it must be accepted that this by definition is not their major priority, which is the education of pupils. Secondly, however progressive school departments or faculties are, they follow their own chosen system, suited to their situations and quite valid in the context of the increasing degree of autonomy from their LEAs that they now experience. None the less, they are on average likely to function as conservative forces. Reliance purely on practical experience can easily result in stagnation and even complacency. Conventional wisdom does not adapt easily to changing situations. Even the most professional of teachers cannot have the time to engage in the wider reading demanded of the higher education teacher. The great advantage of basing teacher education in higher education institutions is that geographical and other educationists have access to a wide range of approaches, as between schools in their areas, revealed as successful in some schools though not in others, through contacts with their peers in other institutions, through working with different LEAs, through reading the relevant range of national and international journals and through the wider research and writing they are given the time and responsibility to engage in. The initial teacher education element in the provision of a quality teaching force for the future remains under threat.

Another potential problem relates to the emergence of a near-minimalist post-Dearing geography curriculum. As has been shown, in its improvements Dearing has potentially sacrificed something of balance and coherence in geography, albeit more at Key Stage 2 than Key Stage 3. There is therefore a test for future teacher professionalism in ensuring that the internal balance and coherence of the subject are not lost. On the positive side, however, teachers are helped by the fact that the teaching approaches advocated in Dearing are, at least in the rhetoric, enquiry based, in so far as they are specified at all, and that teacher choice, within the statutory framework, is not prejudiced.

Finally, if there are 5 years of stability, as Dearing has promised, so long as the queries above are met, there will be the opportunity for the common frameworks of the National Curriculum to provide a basis for improvement by the year 2000.

The Technological Revolution

An inevitable and largely constructive element in the changing context of education is the unprecedented growth in information technology. The changes will continue to stretch the thinking of teachers and students into the twenty-first century. Resource problems of providing the hard- and soft-ware apart, there are undreamt of possibilities in this change. As has often been said, computer technologies have the power to revolutionise teaching. They represent the invention of a genuinely new wheel, in a way that much claimed innovation does not. There are notable developments to come in geographical information systems, in satellite imagery, in interactive video, and the like. The challenge of the micro is being followed by the challenge of CD-ROM, on which the equivalent of 200,000 pages of printed text can be accommodated per disk: a prodigious but daunting tool. Telecomputing networks make regional, national and international networking a practical possibility for schools. Twinning arrangements can be supported by fax and E-mail, for example, offering new opportunities for promoting international understanding.

Not least of the advantages are the quantum leaps achievable in quality production of school-based materials. Computer technology will hopefully play an increasingly emancipatory role in schools in the years towards 2000 and beyond. The use of word processors has enhanced students' presentation of their work, while desk-top publishing in schools has promoted both creativity and useful life-skills. Computer-assisted learning, where properly applied, has materially assisted enquiry-based and problem-oriented activity in school geography. The spatial and environmental foci of geography are more readily exploited by students through the digital manipulation of data, whether in map, graph or visual form.

The National Curriculum in geography demands the practice of IT, as one of a number of cross-curricular skills. It has been variously supported not only by the DFE but by other government departments. The Geographical Association has done much to promote its implementation and will continue to do so. The caveat relates to the potentially negative bandwagon effect, and the possibility that it is the medium, with its rapidly built-in obsolescence factor, rather than the formative quality of the message, which preoccupies the thinking. The enthusiasm for information technology must be 'tempered with critical perspectives', as Kent has urged (1992, p.174). Kent and Philips in fact have indicated that for all the progress, the actual impact of IT in the classroom to date has been fitful and disappointing. In the context of the prevailing 'boosterism', they record the following concerns about the future of IT in schools:

- that children will interact more with machines than with people
- that the IT 'tail' may too readily wag the geography curriculum 'dog'
- that the opportunity costs of IT developments may be too great
- that the vocational relevance of IT programmes will be overplayed
- that educators and their educational objectives will be powerless in the face of the powerful computer/industrial lobby
- that an extension of IT can reinforce power, extending exploitation and control
- that there will be an increasing polarisation between schools with a

richness of IT experience and those in less favourable circumstances
- for gender imbalance in the ways that IT across the curriculum is delivered in schools (1994, p.105).

Futures at the Frontiers: the Implications of Geographical Research

The late 1980s and early 1990s have, as indicated in Chapter 2, witnessed major philosophical changes at the frontiers of the subject. Influential voices have demanded a more distinctive and coherent geography, condemning the fragmentation that both the quantitative and later welfare revolutions brought to the subject. Perhaps the most important trend, as discussed in Chapter 2, is the return of a revived place orientation to the heart of the conceptual frameworking of the subject. As Johnston has argued:

> 'The case for geography developed here is that we must recognise the need to study wholes – places that are milieux...within which ways of life are constructed and reconstructed and within which individuals are socialised into an appreciation of who they are and what is expected of them.' (1991, pp.255–6)

Physical geographers have increasingly focused on establishing links with human and environmental issues (Mannion and Bowlby, 1992), through hazards research and work on environmental design, for example. But the challenges have become greater. Unwin has reiterated the concern that the split of geography into earth and social sciences 'seems likely to leave a void of research on critical issues concerning the human utilisation of the environment' (1992, p.210), and also makes the case for sustaining a distinctive geographical discipline. Supporting Johnston, he concludes:

> 'Place has become a focus for understanding the interaction of the human world of experience and the physical world of existence. The task of a critical geography is to enable people to reflect upon this interaction, and in so doing to create a new and better world.' (1992, p.211)

The return to place assumes that the critical, welfare orientation of geography in the 1970s and 1980s will remain. Johnston's position suggests that through detailed place studies geography can more readily come to grips with the detail of individual human relationships, as distinct from the focus on aggregates with which a substantial part of the research on welfare issues has been preoccupied. Stoddart's conclusion is complementary: that geography must ask the big questions and claim the high ground as the millenium approaches, based on three fundamental qualities: the **real** (based on places), the **unified** (linking human and physical geography) and the **committed** (addressing large scale global issues) (1987, p.333).

As Chapter 2 indicated, the position of radical geography has shifted.

While Marxism and feminism still provide a powerful impetus, there is now a range of Marxisms and feminisms, and a move of radical geography into eco-socialism, offering confusing and even conflicting messages to schools, though not messages that should therefore be disregarded. Ethnic and gender and, hopefully, age-related issues in geography, will rightly loom large in the approach to the year 2000, together with other matters associated with the human rights agenda.

The likely influence of post-modernist thought on geography is difficult to speculate about, offering many contradictions and unclear messages. It has been said, for example, that post-modernist styles are about the life-styles of the young: the cinema, television, rock music, soap opera, dance, and fashion (Gilbert, 1992, p.56), as well as the primacy of personal opinion. Sensitive educators have long had to take on board the tensions between popular and high culture. At the same time, can educators accommodate to what is also defined as an attack on rationality, enlightenment, and the promotion of individual autonomy, in the sense outlined in Chapter 1? Do we accept the view that modernism represents a struggle by an adult establishment to impose a coherent world view on the young? As post-modernist thought in the university context has also been categorised in conspiratorial terms, as symptomatic of academic malaise (Skeggs,1991, p.256), and as the concept is so ill-defined and idiosyncratically used, it is difficult to offer guidance on how it might be accommodated in the school setting. Certainly geography teachers must take on board both the technological revolution that is a tangible manifestation of post-modern society, and such aspects as the time–space compression which are having so powerful an impact on the world and its peoples.

More compatible with the philosophical position of this book than radical and post-modernist thought, is a specifically humanist slant on the purposes we are pursuing. Significantly, the humanist geography literature is not based on Thatcherite qualities: polarisation, transfer of blame, no-u-turn ideology, focus on the aggregate and conspiratorial views of the world, just as, from the opposite end of the political spectrum, some of the radical literature appears to do. Humanist values include the promotion of personal autonomy, respect for the individual, emancipation, regard for human knowledge and understanding, enlightenment values, and also concern for the human condition, social commitment, ecumenicalism, and acceptance of the significance of human agency. In geographical terms, key ideas are the earth as the home of human beings, the former being treated in ecologically sensitive ways, and the latter with empathy. As a stance, humanism:

'welcomes the creative potential of individuals and groups to deal with the surface of the earth in responsible ways. Nor is human creativity confined to the intellectual sphere: it involves emotion, aesthetics, memory, faith and will....It can inspire practitioners of physical, economic, cultural or social geography, and should perhaps not invest too much energy in staking claims for becoming a special branch of the field. Humanism...is leaven in the dough and not a separate loaf in the smorgasbord of geographic endeavour. The Renaissance of humanism calls for an ecumenical rather than a separatist spirit.' (Buttimer, 1993, pp.220–1)

Trends in Geographical Education

The view of Rawling (1993), that in the areas of curriculum thought, research and publication, the 1980s have been a period of impoverishment, is one that is shared here. At the most straightforward level there is the concrete evidence of what those appointed as geographical educationists in the 1960s and 1970s achieved at that time and have achieved since. The late 1960s and 1970s saw them engaged in promotional activity, as described in Chapter 2, in writing methodological texts, and in running major Schools Council and other development projects. There has been a deintellectualisation and diminution of this activity in important respects during the 1980s. With some exceptions, their publications have largely become edited compilations of articles, rather than coherent and substantial methodological monographs. There has been a shift from the attempt to offer in such monographs a balanced synthesis of geographical, educational and social education variables, towards a distinct bias towards the last, as welfare approaches; and a non-place-specific, issues-based orientation has taken over much of secondary school geography, and also initial teacher education and INSET activities in the subject.

The synthesis offered by M.Ed degrees has largely been lost and replaced by a myriad of vocationally oriented short courses. The increasing influence of local education authorities in the late 1970s and during the 1980s radically changed the face of post-experience provision, away from reflective curriculum and other educational theory, towards more concentrated and practically oriented provision. In many respects, the LEA influence has been a positive force, in that work of exceptional quality, dedicated to local area needs, has been promoted, as have the resources to underpin it. It is a matter of some interest and potential value, in that, in contrast to its former divide and rule policy, the government is now actively promoting co-operation between LEAs and teacher education institutions in INSET work.

There are other reasons for the under-achievement of the 1980s in the field of education. For example, many of the geographical educationists appointed in the 1960s and 1970s were later diverted into administrative functions in their departments, and also to other areas of research, which perhaps enjoyed a greater intellectual respectability than applied classroom research in subjects. Others became involved in the political confrontations with government, in the defence of the subject against its possible demise in the mid-1980s. More writing acquired a polemical rather than intellectual thrust as geographical educationists became protagonists against what they perceived as the negative features of the National Curriculum.

Geographical educationists of the early 1970s had been very conscious of developments at the frontiers of the subject, and there were strong links established in many cases between them and their colleagues in subject departments, to which the earlier account of the mutual implementation of the quantitative revolution testifies. As was complained of in Chapter 2, much less attention has been paid by those in education to the important developments at the frontiers of the subject in the late 1980s and early 1990s.

Thus hopes for the future lie in re-establishing these links. It is fortunate that there has been sufficient convergence towards the view that place and space are central to a distinctive geography at both the academic frontiers and at National Curriculum levels, to make constructive linkages possible. My other hope, though perhaps less likely to be achieved, is that future thinkers will take more account of the cultural heritage they have inherited than hitherto. Thus we continue to present the past of geographical education as an annal of progress from the capes and bays primitivism, through the arid regional paradigm, to the higher achievements of the present. We continue to purvey negative stereotypes of the past, including the recent past. It is salutary to recall that enquiry-based field work has been advocated at least since the 1840s, and focused locality-type studies from at least the 1890s. The past offers us many elements to be conserved, as well as some to be rejected.

Trends in Social Education

In the balanced synthesis in curriculum development in geography which this book has attempted to underwrite, the social education variable is difficult to dissociate from the other two. The argument has been that it was this dominating variable that tended to nudge the others out of balance during the 1980s. The official thinking behind the National Curriculum has tried to halt its advance and to replace it with a different and even more unbalanced framework. A promising way forward, as outlined in Chapter 9, is through taking account of what is indeed a balanced approach, namely that of the International Charter on Geographical Education, accepted in 1992 by the International Geographical Union as a global guidline, translated into many different languages.

In his Presidential address to the Geographical Association in 1993, Catling worked some of these IGU recommendations into an agenda for geographers of the future to support. In its central function of furthering the study and teaching of geography, he urged the need for the Association to:

- work both nationally to extend and internationally to achieve geographical study for all children, i.e. to ensure the **right to a geographical education**;
- foster a geography that **supports the development of geographical knowledge, understanding and skills**, seen as a basis for active community involvement from the time of entry to school;
- develop a geography which sees at its centre **values and attitudes founded on international understanding, environmental ethics, justice and human rights**;
- promote the study of the **geography of children**, and the way in geographical processes and environmental and political issues impact on their lives and opportunities in different places;
- work for a **geography for action**, which is founded on children's active participation in community decision making (1993, p.356).

The agenda is summarised as being one for geography **about** the community and environment, a geography **in** the community and environment, and a geography **for** the community and environment.

There is no quarrel in principle here with the proposed agenda. The cautionary note must be about its balance with other priorities for education, i.e. geographical and pedagogic variables, and about the processes by which it can be achieved. It must never be forgotten that politically active groups have perennially used different methods to achieve their ends, some of them educationally and socially indefensible. Some animal welfare groups bring to bear politically legitimate pressures, for example; others fascist methods of browbeating and violating their opponents. Thus an approach to an environmental issue must be different when confronting counter-interests on a political platform, than when trying to educate children about the issue. As argued previously, an important approach to human rights education is, rather than starting by teaching for an awareness and understanding of global aggregates related to human rights, to deal with similar issues at individual and group human relations levels, through studies of particular places. That is why the revival of the place approach, into which issues are permeated, seems to have so much to offer for the future.

In the social education leading to the millenium, students need to have a voice, with a more future-oriented perspective built into geography programmes. In a pamphlet on *Education for the 21st Century*, the World Wildlife Fund (1993) have offered guidelines on educating for the future, which are relevant to geography teachers at all phases:

- develop a more future-oriented perspective on pupil's own lives and events in the wider world

- identify and envision alternative futures that are more just and sustainable
- exercise their critical thinking skills and creative imagination more effectively
- participate in more thoughtful and informed decision making in the present
- engage in active and responsible citizenship in the local and global community, on behalf of present and future generations.

Curriculum Development Trends

Curriculum development in the future, I would argue, should seek to build on the following:

- as suggested above, re-establishing links with:
  - the frontiers of geography
  - a revived curriculum theory
- a greater recognition of both the distinctiveness of, as well as common-alities between, subjects
- a recognition that the issues with which subjects such as geography have to grapple are essentially inter-disciplinary, and must be addressed both through geography's distinctive contributions, and from a whole curriculum perspective and, following from this
- a greater awareness of what is happening in other parts of the curriculum, as a necessary background to evolving more effective subject-based topic work
- more co-operation between higher education and local education authorities in INSET provision.

As always, any favourable judgement on the value of geography in the curriculum, as for other subjects, must be accompanied by the qualification 'if well taught'. This implies not only technical and human skills, but also a balanced framework embodying the critical educational variables of infusing meaning and refining thinking, as well as of seeking a major role for the subject in striving to achieve more sustainable environmental resource use and a better quality of life for the world's citizens. If the balance between these variables is not even being sought, however, then 'I am quite willing to admit', as Ravenstein once put it, 'that the hours spent upon geographical instruction might be employed to better purpose' (1885, p.163).

References

Agnew, J.A. and Duncan, J.S. (eds) (1989) *The Power of Place: Bringing Together Geographical and Sociological Imaginations*. Boston: Unwin Hyman.

Ahier, J. (1988) *Industry, Children and the Nation: an Analysis of National Identity in School Textbooks*. Lewes: The Falmer Press.

Alexander, R. (1984) *Primary Teaching*. London: Holt, Rinehart and Winston.

Alexander, R. (1994) 'The classteacher and the curriculum', in Pollard, A. and Bourne, J. (eds) *Teaching and Learning in the Primary School*. London: Routledge, in association with the Open University, pp. 206–12 .

Ambrose, P.J. (1969) 'Perceptions of space, distance and the environment', in Ambrose, P.J. (ed.) (1969) *Analytical Human Geography*. London: Longman, pp.172–4.

Anonymous. (1918) 'Children of other lands: Spain', *The Schoolmistress*, 23 May 1918, p.116.

Arnold, M. (1867) 'General report for 1867', in Board of Education (1910) *Reports on Elementary Schools 1852–1882 by Matthew Arnold*. London: HMSO .

Arnold, M. (1869) 'General report for 1869', in *ibid.*, pp.129–30.

Atkins, E.J. (1988) 'Reframing curriculum theory in terms of interpretation and practice: a hermeneutical approach', *Journal of Curriculum Studies*, 20, 437–48 .

Ausubel, D.P. and Robinson, F.G. (1969) *School Learning: an Introduction to Educational Psychology*. New York: Holt, Rinehart and Winston.

Bailey, P. (1974) *Teaching Geography*. Newton Abbot: David and Charles.

Bailey, P. (1989) 'A place in the sun: the role of the Geographical Association in establishing geography in the National Curriculum of England and Wales', *Journal of Geography in Higher Education*, 13, 149–157.

Bailey, P. and Binns, T. (eds) (1987) *A Case for Geography*. Sheffield: The Geographical Association.

Baker, A.R.H. and Biger, G. (eds) (1992) *Ideology and Landscape in Historical Perspective: Essays on the Meanings of Some Places in the Past*. Cambridge: Cambridge University Press .

Baker, A.R.H. (1992) 'Introduction: on ideology and landscape', in *ibid.*, pp. 1–14.

Bale, J. (1994) 'Geography teaching, postmodernism and the National Curriculum', in Walford, R. and Machon, P. (eds) *Challenging Times: Implementing the National Curriculum in Geography*. Cambridge: Cambridge Publishing Services, pp. 95–7.

Bale, J. (1995) 'Postmodernism and geographical education', in Williams, M. (ed.) *Understanding Geographical and Environmental Education: the Role of Research*. London: Cassell.

Bale, J., Graves. N.J. and Walford, R. (eds) (1973) *Perspectives on Geographical Education*. Edinburgh: Oliver,and Boyd.

Ball, N. (1964) 'Richard Dawes and the teaching of common things', *Educational Review*, 17, 62–4.

Ballard, P.B. (1923) *The New Examiner*. London: University of London Press, (1949 edn.)

Barke, M. and O'Hare, G. (1984) *The Third World: Diversity, Change and Interdependence*. Edinburgh: Oliver and Boyd.

Barnes, D. (1969) 'Language in the secondary classroom', in Barnes, D. *et al. Language, the Learner and the School*. Harmondsworth: Penguin, pp. 11–77.

Beddis, R. and Mares, C. (1988) *School Links International: a New Approach to Primary School Linking round the World*. Bristol: Avon County Council/Brighton: Tidy Britain Group.

Bennetts, T. (1994) 'The draft proposals for geography: a personal response', *Teaching*

Geography, **19**, 160–1.

Bernstein, B. (1971) 'On the classification and framing of educational knowledge', in Young, M.F.D. (ed.) *Knowledge and Control: New Directions for the Sociology of Education.* London: Collier-Macmillan, pp. 47–69.

Bird, J. (1989) *The Changing Worlds of Geography: a Critical Guide to Concepts and Methods.* Oxford: Clarendon Press.

Bloom, B.S. (ed.) (1956) *Taxonomy of Educational Objectives: Handbook I: Cognitive Domain.* London: Longman, 1972 edition, pp. 201–7.

Bloom, B.S. (1968) 'Mastery learning', in Block , J.H. (ed.) (1971) *Mastery Learning: Theory and Practice.* New York: Holt, Rinehart and Winston, pp.447–63.

Bloom, B.S., Hastings, J.T. and Madaus, G.F. (1971) *Handbook on Formative and Summative Evaluation of Student Learning.* New York: McGraw Hill .

Blyth, W.A.L., Cooper, K., Derricott, R., Elliott, G., Sumner, H. and Waplington, A. (1972) *History, Geography and Social Science 8–13: an Interim Statement.* London: Schools Council.

Blyth, W.A.L., Copper, K., Derricott, R., Elliott, G., Sumner, H. and Waplington, A. (1976) *Place, Time and Society 8–13: Curriculum Planning in History, Geography and Social Science.* Bristol: Collins/ESL for the Schools Council.

Board of Education (1905) 'The Teaching of Geography', in *Suggestions for the Consideration of Teachers and others Concerned in the Work of Public Elementary Schools.* London: HMSO, pp.54–60.

Board of Education (1914) 'Suggestions for the teaching of geography', in *Suggestions for the Consideration of Teachers and others Concerned with the Work of Public Elementary Schools.* London: HMSO .

Board of Education (1931) *Report of the Consultative Committee on the Primary School (Hadow Report).* London: HMSO.

Boardman, D. (1982) 'Assessing readability', in Boardman, D. (ed.) (1982) *Geography with Slow Learners.* Sheffield: The Geographical Association, pp.67–72.

Boardman, D. (1983) *Graphicacy and Geography Teaching.* London: Croom Helm.

Boardman, D. (ed.) (1986) *Handbook for Geography Teachers.* Sheffield: The Geographical Association.

Boardman, D. (1988) *The Impact of a Curriculum Project: Geography for the Young School Leaver.* Birmingham: The University of Birmingham.

Bobbitt, J.F. (1918) *The Curriculum.* Boston: Houghton Mifflin/ New York: Arno Press reprint, 1971.

Boden, P. (1976) *Developments in Geography Teaching.* London: Open Books.

Booth, J. (1857) *How to Learn and What to Learn.* London: Bell and Daldy.

Bowlby, S. (1992) 'Feminist geography and the changing curriculum', *Geography,* **77**, 349–60.

Bowlby, S. *et. al.* (1988) *Gender and Geography: Contemporary Issues in Geography and Education,* Vol. 3 (1). London: Association for Curriculum Development, pp. 1–7.

Bradford, M.G. and Kent, A. (1977) *Human Geography: Theories and their Applications.* Oxford: Oxford University Press.

Briggs, K. (1972) *Introducing Transportation Networks.* London: University of London Press.

Bristow, E.J. (1977) *Vice and Vigilance: Purity Movements in Britain since 1700.* Dublin: Gill and Macmillan.

Bruner, J.S. (1960) *The Process of Education.* New York: Random House.

Bruner, J.S. (1966) *Toward a Theory of Instruction.* Cambridge, Mass: Harvard University Press.

Butt, G. (1994) 'Geography, vocational education and assessment', *Teaching Geography,* **19**, 182–3.

Buttimer, A. (1993) *Geography and the Human Spirit.* Baltimore: Johns Hopkins University Press.

Carnie, J. (1966) 'The development of national concepts in junior school children', in Bale, J., *et al.* (eds) (1973), pp.101–18.

Carroll, J.B.(1963) 'A model of school learning', *Teachers College Record,* **64**, 723–33 .

Carswell, R.J.B. (1970) 'Evaluation of affective learning in geographical education', in Kurfman, D.G. (ed.) *Evaluation in Geographic Education.* Belmont, California: Fearon Publishers.

Carter, C.C. (1961) *Landforms and Life.* London: Christophers.

Carter, R. (1991a) 'A matter of values', *Teaching Geography,* **16**, 30.

Carter, R. (1991b) *Talking about Geography: the Work of Geography Teachers in the National Oracy Project.* Sheffield: The Geographical Association .

Carter, R. (1994) 'The GA responds to the draft proposals for geography', *Teaching Geography,* **19**, 158–9.

Catling, S. (1988) 'On the move', in C.S. Tann, *Developing Topic Work in the Primary School.*

Lewes: The Falmer Press, pp. 149–65.

Catling, S. (1993) 'The whole world in our hands', *Geography*, **78**, 340–58.

Centre for Educational Research and Innovation (CERI) (1972) *The Nature of the Curriculum for the 1980s and Onwards*. Paris: OECD, p.61.

Child Life (1900) 'Children and the war', **2**, 50–1.

Chorley, R. and Haggett, P. (eds) (1965) *Frontiers in Geographical Teaching*. London: Methuen.

Cole, J.P. and Beynon, N.J. (1968ff.) *New Ways in Geography* series: Oxford: Basil Blackwell.

Cons, C.J. and Fletcher, C. (1938) *Actuality in School: an Experiment in Social Education*. London: Methuen.

Cooke, P. (ed.) (1989) *Localities: the Changing Face of Urban Britain*. London: Unwin Hyman.

Corbridge, S. (1989) 'Debt, the nation state, and theories of the world economy', in Gregory, D and Walford, R. (eds) *Horizons in Human Geography*. London: Macmillan, pp. 341–60.

Corney, G. and Rawling, E. (eds) (1985) *Teaching Slow Learners through Geography*. Sheffield: The Geographical Association.

Coulthard, E.M. (1938) Contribution on 'Local studies' to the Geographical Association Steering Committee for Geography in Secondary Schools, *Geography*, **21**, 178.

Coulthard, E.M. (1946) 'A spring term experiment', *Geography*, **31**, 134–8.

Courtis, S.A. (1930) 'Individualisation in education', *University of Michigan School of Education Bulletin*, **1**, 52.

Cowham, J.H. (1900) *The School Journey: a Means of Teaching Geography, Physiography and Elementary Science*. London: Westminster School Book Depot.

Daniels, S. (1992) 'Place and the geographical imagination', *Geography*, **77**, 310–22.

Daugherty, R. (1990) 'Assessment in the geography curriculum', *Geography*, **75**, 289–301.

Daugherty, R. and Lambert, D. (1994) 'Teacher assessment and geography in the National Curriculum', *Geography*, **79**, 339–49.

Davis, W.M. (1902) 'Field work in physical geography', in Johnson, D.W. (1909) *Geographical Essays by William Morris Davis*. London: Constable, (1954 edn.), pp.236–48.

Department of Education and Science (1978a) *Primary Education in England: a Survey by HM Inspectors of Schools*. London: HMSO.

Department of Education and Science (1978b) *The Teaching of Ideas in Geography: some Suggestions for the Middle and Secondary Years of Schooling: a Discussion Paper by some HM Inspectorate of Schools*. London: HMSO.

Department of Education and Science (1986) *Geography from 5–16. Curriculum Matters 7*. London: HMSO.

Department of Education and Science (1981) *The School Curriculum*. London: HMSO.

Department of Education and Science (1989a) *Aspects of Primary Education: the Teaching and Learning of History and Geography*. London: HMSO .

Department of Education and Science/Welsh Office (1989b) *National Curriculum Geography Working Group: Interim Report*. London: DES.

Department of Education/Welsh Office (1990) *Geography for Ages 5–16: Proposals of the Secretary of State for Education and Science and Secretary of State for Wales:* London:DES.

Department of Education and Science (1991) *National Curriculum: Draft Order for Geography*. London: DES.

Department for Education (1992) *Policy Models: a Guide to Developing and Implementing European Dimension Policies in LEAs, Schools and Colleges*. London: DfE.

Development Education Centre (1986) *Theme Work: Approaches for Teaching with a Global Perspective*. Birmingham: Development Education Centre.

Dewey, J. (1902) *The Child and the Curriculum*. Chicago: Phoenix Books, University of Chicago Press.

Dewey, J. (1916) *Democracy and Education: an Introduction to the Philosophy of Education*. New York: The Free Press .

Digby, B. (1994) 'Stranded! reflections upon assessment at Key Stage 3', in Walford R. and Machon, P. (eds), *Challenging Times: Implementing the National Curriculum in Geography*. Cambridge: Cambridge Publishing Services, pp.80–3.

Dilkes, J.L. and Nicholls, A. (eds) (1988) *Low Attainers and the Teaching of Geography*. Sheffield: The Geographical Association/Stafford: The National Association for Remedial Education.

Doll, W.E. (1993) *A Post–modern Perspective on Curriculum*. New York: Teachers College Press.

Downs, R.M. and Stea, D. (1973) 'Cognitive maps and spatial behaviour: process and products', in Downs, R.M. and Stea, D. (eds) *Image and Environment: Cognitive Maps and Spatial Behaviour*. London: Edward Arnold, pp. 8–26.

Downs, R.M. and Stea, D. (1977) *Maps in Minds: Reflections on Cognitive Mapping*. New York: Harper and Row.

Entwistle, H. (1971) 'The relationship between theory and practice', in Tibble, J.W. (ed.) *An Introduction to the Study of Education: an Outline for the Student*. London: Routledge and Kegan Paul, pp.95–113.

Evans, C. (1933) 'Geography and world citizenship' in *The Teaching of Geography in Relation to the World Community*, The School and the Wold Community Series, Pamphlet 1, Cambridge: Cambridge University Press, pp. 16–19.

Everson, J. and Fitzgerald, B.P. (1969) *Settlement Patterns*. London: Longman.

Eyles, J. (1989) 'The geography of everyday life', in Gregory and Walford (eds), *Horizons in Human Geography*. London: Macmillan, pp. 102–17.

Eysenck, H.J. (1973) 'A better understanding of IQ and the educational myths surrounding it', *Times Educational Supplement*, **18 May**, 2.

Fairgrieve, J. (1924) *Geography and World Power*. London: University of London Press (1932 reprint).

Fairgrieve, J. and Young, E. (1939) *Real Geography: Book 1*. London: G. Philip and Son.

Fien, J. and Gerber, R. (eds) (1988) *Teaching Geography for a Better World*. Edinburgh: Oliver and Boyd.

Fien, J. and Slater, F. (1985) 'Four strategies for values education in geography', in Boardman, D. (1985) *New Directions in Geographical Education*. Lewes: The Falmer Press, pp.171–86.

Finch, P.J. (1931) *Geography through the Shop Window*. London: Evans Brothers.

Fisher, S. and Hicks, D. (1985) *World Studies 8–13: a Teacher's Handbook*. Edinburgh: Oliver and Boyd.

Fishman, J.A. (1956) 'An examination of the process and function of social stereotyping', *The Journal of Social Psychology*, **43**, 31–5.

Fitch, J.G. (1884) *Lectures on Teaching*. Cambridge: Cambridge University Press.

Fitzgerald, B.P. (1974) *Developments in Geographical Method*, Vol. 1 in *Science in Geography* series. Oxford: Oxford University Press.

Forsaith, D.M. (1951) *Many People in Many Lands*. London: University of London Press.

Freshfield, D.W. (1886) 'The place of geography in education', *Proceedings of the Royal Geographical Society*, New Series, **8**, 698–718.

Fry, P. and Schofield, A. (eds) (1993) *Geography at Key Stage 3: Teachers' Experience of National Curriculum Geography in Year 7*. Sheffield: The Geographical Association .

Gagne, R.M. (1965) *The Conditions of Learning*. New York: Holt, Rinehart and Winston.

Gagne, R.M. (1967) 'Learning theory, educational media, and individualized instruction', in Hooper, R. (ed.) (1971) *The Curriculum: Context, Design and Development*. Edinburgh: Oliver and Boyd, pp. 299–319.

Garnett, A. (1969) 'Teaching geography: some reflections', *Geography*, **54**, 385–400.

Garnett, O. (1940) ' Reality in geography', *The Journal of Education*, **72**, 171–3.

Geikie, A. (1879) 'Geographical evolution', *Proceedings of the Royal Geographical Society*, New Series, **1**, 422–43.

Geikie, A. (1882) 'My first geological excursion', in Geikie, A. *Geological Sketches at Home and Abroad*. London: Macmillan and Co., pp.1–25.

Geikie, A. (1887) *The Teaching of Geography*. London: Macmillan and Co. (1908 edn.).

Geographical Association (1934) 'Memorandum for the Consultative Committee of the Board of Education', *Geography*, **20**, 47–51.

Geographical Association (1962) *Sample Studies*. Sheffield: The Geographical Association.

Geographical Association (1968) *Asian Sample Studies*. Sheffield: The Geographical Association.

Gilbert, R. (1992) 'Citizenship, education and post–modernity', *British Journal of Sociology of Education*, **13**, 51–68.

Gipps, C. and Murphy, P. (1994) *A Fair Test? Assessment, Achievement and Equity*. Buckingham: Open University Press, pp. 29–64.

Goldsmith, J. (1813) *An Easy Grammar of Geography for Schools and Young Persons*. London: Longman, Hunt, Rees, Orme and Brown..

Goldsmith, J. (1823) *A Grammar of General Geography*. London: Longman, Hunt, Rees, Orme and Brown.

Goodall, S. (ed.) (1994) *Developing Environmental Education in the Curriculum*. London: David Fulton Publishers.

Goodson, I.F. (1983) *School Subjects and Curriculum Change: Case Studies in Curriculum History*. London: Croom Helm, pp. 60–88 .

Gould, P. and White, R. (1974) *Mental Maps*. Harmondsworth: Penguin Books.

Graham, D. with Tytler, D. (1993) *A Lesson for us All: the Making of the National Curriculum*. London: Routledge.

Graves, N.J. (1968) 'Geography, social science and inter–disciplinary enquiry', *Geographical*

220

Journal, **134**, 390–4.

Graves, N.J. (ed.) (1972) *New Movements in the Study and Teaching of Geography*. London: Temple Smith.

Graves, N.J. (1975) *Geography in Education*. London: Heinemann.

Graves, N.J. (1979) *Curriculum Planning in Geography*. London: Heinemann .

Graves, N.J.and Naish, M. (eds) (1986) *Profiling in Geography*. Sheffield: The Geographical Association.

Green, A. (1994) 'Postmodernism and state education', *Journal of Educational Policy*, **9**, 67–83.

Gregory, D. (1978) *Ideology, Science and Human Geography*. London: Hutchinson.

Gregory, D. and Walford, R. (1989) *Horizons in Human Geography*. London: Macmillan.

Gregory, D. (1989) 'Foreword', in Kobayashi, A. and Mackenzie, S. (eds) *Remaking Human Geography*. Boston: Unwin Hyman, p.v

Griffiths, V.L. and Abdel Rahman ali Taha (1939) *Beginning Geography in Africa and Elsewhere*. London: Evans.

Haggett, P. (1972) *Geography: a Modern Synthesis*. New York: Harper and Row.

Haggett, P. and Chorley, R. (1989) 'From Madingley to Oxford: Foreword to *Remodelling Geography*', in Macmillan, B. (ed.) Remodelling Geography. Oxford: Blackwell, pp. xv–xx.

Hall, D. (1976) *Geography and the Geography Teacher*. London: George Allen and Unwin.

Hall, G. (1989) *Records of Achievement: Issues and Practice*. London: Kogan Page.

Hall, G. (ed.) (1992) *Themes and Dimensions of the National Curriculum: Implications for Policy and Practice*. London: Kogan Page.

Hamilton, W.R. (1838) 'Address', *Journal of the Royal Geographical Society of London*, **8**, xxxix–xl.

Hansard (1861), COI 2111.

Hargreaves, A. (1982) 'The rhetoric of school–centred innovation', *Journal of Curriculum Studies*, **14**, 251–66 .

Harlen, W., Gipps, C., Broadfoot, P. and Nuttall, D. (1994) 'Assessment and the improvement of education', in Pollard, A. and Bourne, J. (eds) *Teaching and Learning in the Primary School*. London: Routledge, pp. 219–27.

Harris, A. (1972) 'Autonomy' in Jenkins, D., Pring, R. and Harris, A. (eds) *Curriculum Philosophy and Design*. Milton Keynes: Open University Press, Unit 8..

Harris, B.F.D. (1934) 'The claim of geography to be considered as a science, and implications as to methods of teaching the subject', *Geography*, **20**, 38–46.

Hart, R.A. and Moore, G.T. (1973) 'The development of spatial cognition: a review', in Downs, R.M. and Stea, D. (eds) *Maps in Minds: Reflections on Cognitive Mapping*. New York: Harper and Row, pp. 246–88.

Harvey, D. (1969) *Explanation in Geography*. London: Edward Arnold.

Harvey, D. (1973) *Social Justice and the City*. London: Edward Arnold.

Harvey, D. (1989) *The Condition of Postmodernity: an Enquiry into the Origins of Cultural Change*. Oxford: Basil Blackwell.

Harvey, D. (1989) 'From models to Marx: notes on the project to "remodel" contemporary geography', in Macmillan, B. (ed.) *Remodelling Geography*. Oxford: Blackwell, pp.211–16 .

Helburn, N. (1968) 'The educational objectives of High School geography', *Journal of Geography*, **67**, 280–2.

Herbertson, A.J. (1896) 'Geographical education', *Scottish Geographical Magazine*, **12**, 414–21.

Herbertson, A.J. and Herbertson, F.D. (1899) *Man and his Work: an Introduction to Human Geography*. London: A. and C. Black.

Herbertson, A.J. (1905) 'The major natural regions of the world', *Geographical Journal*, **25**, 300–10.

Hickman, G.M. (1950) 'The sample study – a method and its development', *Journal of Geography*, **49**, 151–9.

Hickman, G.M., Reynolds, J. and Tolley, H. (1973) *A New Professionalism for a Changing Geography*. London: Schools Council.

Hicks, D. (1981) 'Images of the world: what do geography text–books actually teach about development?', *Cambridge Journal of Education*, **11**, 15–35.

Hicks, D. and Steiner, M. (eds) (1989) *Making Global Connections: a World Studies Workbook*. Edinburgh: Oliver and Boyd.

Hirsch, E.D. (1987) *Cultural Literacy: What Every American Needs to Know*. Boston: Houghton Mifflin Co.

Holly, D. (1973) *Beyond Curriculum*. London: Hart–Davis MacGibbon.

Holmes, E. (1911) *What Is and What Might Be: a Study of Education in General and Elementary Education in Particular*. London: Constable.

Honeybone, R.C. (1954) 'Balance in geography and education', *Geography*, **39**, 90–101 .

Honeybone, R.C. (ed.) (1956ff.) *Geography in School* series: London: Heinemann.

Houston, J.F. (1952) *South Africa.* Edinburgh: Oliver and Boyd.

Howarth, V. B. R. (1954) 'The Commonwealth in the geography syllabus', *Geography 39*, 5–12.

Huckle, J. (ed.) (1983) *Geographical Education: Reflection and Action.* Oxford: Oxford University Press.

Hughes, J. and Thomas, D. (1994) 'Geography for special children', in Marsden, W.E. and Hughes, J. (eds) *Primary School Geography.* London: David Fulton, pp.170–85.

Hurt, J.S. (1977) 'Drill, discipline and the elementary school ethos', in McCann, W.P.(ed.) *Popular Education and Socialization in the Nineteenth Century.* London: Methuen, pp.167–91.

Huxley, T.H. (1877) *Physiography: an Introduction to the Study of Nature.* London: Macmillan and Co..

International Geographical Union (1992) *International Charter in Geographical Education.* Commission in Geographical Education.

Jahoda, G. (1963) 'The development of children's ideas and attitudes about other countries', *British Journal of Educational Psychology*, **33**, 47–60.

James, C. (1968) *Young Lives at Stake.* London: Collins.

Johnston, R.J. (1989) 'Philosophy, ideology and geography', in Gregory, D. and Walford, R. (eds) Horizons in Human Geography. London: Macmillan, pp. 48–65.

Johnston, R.J. (1991) *A Question of Place: Exploring the Practice of Human Geography.* Oxford: Blackwell.

Johnston, R.J. and Taylor, P.J. (1986) *A World in Crisis? Geographical Perspectives.* Oxford: Basil Blackwell.

Johnston, R.J., Hauer, J. and Hoekveld, G.A. (eds) (1990) *Regional Geography: Current Developments and Future Prospects.* London: Routledge.

Joseph, K. (1985) 'Geography in the school curriculum', *Geography*, **70**, 290–7.

Keane, A.H. (1920) *Man: Past and Present.* Cambridge: Cambridge University Press.

Keddie, N. (1971) 'Classroom knowledge', in Young, M.F.D. (ed.) (1971) *Knowledge and Control: New Directions for the Sociology of Education.* London: Collier-Macmillan, pp.133–60.

Kelly, A.V. (1986) *Knowledge and Curriculum Planning.* London: Harper and Row.

Kennedy, W.J. (1849) 'Report', *Minutes of the Committee of Council on Education (1849–50).* London: HMSO.

Kent, W.A. (1992) The new technology and geographical education', in Naish, M. (ed.) *Geography and Education: National and International Perspectives.* London: University of London Institute of Education, pp.163–76.

Kent, W.A. and Philips, A (1994) in Marsden, W.E. and Hughes, J. (eds) *Primary School Geography.* London: David Fulton, pp.93–106.

King, W.A. (1838) 'Physiology as connected with education', in Central Society of Education, *2nd Publication.* London: The Woburn Press, 1968 reprint, pp.163–86.

Kingston, W.H.G. (1849) *A Lecture on Colonization.* London..

Kingston, W.H.G. (1871) *In the Wilds of Africa.* London: Nelson.

Kirk, W. (1963) 'The problems of geography', *Geography*, **48**, 357–71.

Kirk, D. (1988) 'Ideology and school-centred innovation: a case study and a critique', *Journal of Curriculum Studies*, **20**, 449–64.

Kobayashi, A. and Mackenzie, S. (eds) (1989) *Remaking Human Geography.* Boston: Unwin Hyman.

Knight, P. (1993) *Primary Geography, Primary History.* London: David Fulton Publishers.

Kohlberg, L. (1976) *Recent Research in Moral Development.* New York: Holt Rinehart and Winston.

Kratwohl, D.R., Bloom, B.S. and Masia, B.B. (1964) *Taxonomy of Educational Objectives: Handbook II: Affective Domain.* London: Longman (1971 edn.).

Kuhn, T.S. (1962) *The Structure of Scientific Revolutions.* Chicago: University of Chicago Press.

Laidler, F.F. (1946) 'Geography in the modern school', *Geography*, **31**, 146–8.

Lambert, D. (1990) *Geography Assessment: a Guide and Resource for Teachers.* Cambridge: Cambridge University Press.

Langton, J. (1972) 'Potentialities and problems of adopting a systems approach to the study of change in human geography', in Board. C., *et al.* (eds) *Progress in Geography*, Vol.4. London: Edward Arnold.

Larsen, B. (1988) 'Gender bias and the GCSE' in Whatmore, S. and Little, J. (eds) (1988) 'Gender and geography', *Contemporary Issues in Geography and Education*, **3**, 80–4.

Law, C.M. and Warnes, A.M. (1976) 'The changing geography of the elderly in England and

Wales', *Transactions of the Institute of British Geographers,* 1 (New Series), 453–71; *see also* Warnes, A.M. and Law, C.M. (1984) 'The elderly population of Great Britain: locational trends and policy implications', *Transactions of the Institute of British Geographers,* 9 (New Series), 37–59 .

Ley, D. (1989) 'Modernism, post–modernism and the struggle for place', in Agnew, J.A. and Duncan, J.S. (eds) *The Power of Place.* Boston: Unwin Hyman, pp. 44–65.

Livingstone, D.N. (1992) *The Geographical Tradition.* Oxford: Blackwell.

Long, E. (1774) *A History of Jamaica.* London.

Long, M. and Roberson, B.S. (1966) *Teaching Geography.* London: Heinemann.

Lowe, R. (1861) in *Hansard,* 165, Col.211.

Lyde, L.W. (1912) 'Discussion on the homework of the geography class', *The Geographical Teacher,* 6, 200–1.

Machon, P. (1991) 'Subject or citizen?', *Teaching Geography,* 16, 128.

Macdonald, B. and Walker, R. (1976) *Changing the Curriculum.* London: Open Books.

McDonald, E.H.B. and Dalrymple, J. (1910) *Little People Everywhere Series: Ume San in Japan.* London: Wells Gardner, Darton and Co.

McDowell, L. (1989) 'Women, gender and the organisation of space', in Gregory, D. and Walford, R. (eds), *Horizons in Human Geography.* London: Macmillan, pp.136–51.

McKewan, N. (1986) 'Phenomenology and the curriculum: the case of secondary school geography', in Taylor, P.H. (ed.) *Recent Developments in Curriculum Studies.* London: NFER: Nelson, pp.156–67.

McKiernan, D. (1993) 'History in the National Curriculum: imagining the nation at the end of the 20th century', *Journal of Curriculum Studies,* 25, 33–51.

Mackinder,H.J. (1887) 'On the scope and methods of geography', *Proceedings of the Royal Geographical Society,* New Series, 9, 141–73.

Mackinder, H.J. (1911) 'The teaching of geography from the imperial point of view and the use which could and should be made of visual instruction', *The Geographical Teacher,* 6, 79–86.

Mackinder, H.J. (1921) 'Geography as a pivotal subject in education', *Geographical Journal,* 57, 376–84.

Mackinder, H.J. (1943) 'The development of geography', *Geography,* 28, 69–77.

MacMunn, N. (1926) *The Child's Path to Freedom.* London: J. Curwen and Sons.

McMurry, C.A. (1899) 'A course of geography for the grades of the common school', *Fourth Yearbook of the National Herbart Society, for 1898, Supplement.* Chicago: University of Chicago Press.

Madge, C. (1994) '"Gendering space": a first year geography fieldwork exercise', *Geography,* 79, 330–8.

Mager, R.F. (1962) *Preparing Instructional Objectives.* Belmont, California: Fearon Publishers.

Mannion, A.M. and Bowlby, S.R. (eds) (1992) *Environmental Issues in the 1990s.* Chichester: John Wiley and Sons.

Manson, G. (1973) 'Classroom questioning for geography teachers', *Journal of Geography,* 72, 24–30.

Mapother, E.D. (1870) *The Body and its Health.* Dublin: John Falconer.

Marland, M. (1973) 'Preference shares', *Education Guardian,* 11 September 1973, p.18.

Marsden, W.E. (1976a) *Evaluating the Geography Curriculum.* Edinburgh: Oliver and Boyd.

Marsden, W.E. (1976b) 'Stereotyping and third world geography', *Teaching Geography,* 1, 228–30.

Marsden, W.E. (1979) 'The language of the geography textbook: an historical appraisal', *Westminster Studies in Education,* 2, 53–65.

Marsden, W.E. (1986) 'The Royal Geographical Society and geography in secondary education', in Price, M.H. (ed.) *The Development of the Secondary Curriculum.* London: Croom Helm, pp. 182–213.

Marsden, W.E. (1988) 'The age aspect in human rights geography in school', *Teaching Geography,* 13, 146–8.

Marsden, W.E. (1989) '"All in a good cause": geography, history and the politicization of the curriculum in nineteenth and twentieth century England ', *Journal of Curriculum Studies,* 21, 509–26.

Marsden, W.E. (1990) 'Rooting racism into the educational experience of childhood and youth in the nineteenth– and twentieth–centuries', *History of Education,* 19, 333–53.

Marsden, W.E. (1991) '"The structure of omission": British curriculum predicaments and false charts of American experience', *Compare,* 21, 5–25.

Martin, G.C. and Wheeler, K. (eds) (1975) *Insights into Environmental Education.* Edinburgh: Oliver and Boyd.

Martland, J. (1994) 'New thinking in mapping', in Marsden, W.E. and Hughes, J. (eds) *Primary School Geography.* London: David Fulton, pp. 37–49.

Masterton, T. (1969) *Environmental Studies*. Edinburgh: Oliver and Boyd.

Matthews, M.H. (1984) 'Cognitive mapping abilities of young girls and boys', *Geography*, **69**, 327–36.

Matthews, M.H. (1986) 'Gender, graphicacy and geography', *Educational Review*, **38**, 259–71.

Matthews, M.H. (1992) *Making Sense of Place: Children's Understanding of Large Scale Environments*. Hemel Hempstead: Harvester Wheatsheaf.

Milburn, D. (1972) 'Children's vocabulary', in Graves, N.J. (ed.) *New Movements in the Study and Teaching of Geography*. London: Heinemann, pp. 107–20.

Mills, D. (ed.) (1988) *Geographical Work in Primary and Middle Schools*. Sheffield: The Geographical Association.

Morgan, W. (1994) 'Making a place for geography: the Geographical Association's initiatives and the Geography Working Group's experience', in Marsden, W.E. and Hughes, S. (eds) *Primary School Geography*. London: David Fulton, pp.23–36.

Musgrove, F. (1973) 'Power and the integrated curriculum', *Journal of Curriculum Studies*, **5**, 3–12.

Naish, M. (1985) 'Geography 16–19', in Boardman, D. (ed.) *New Directions in Geographical Education*. Lewes: The Falmer Press, pp. 99–115.

Naish, M. (1991) 'Geography in the secondary curriculum', in Naish, M. (ed.) *Geography and Secondary Education: National and International Prospectives*. London: University of Education, pp. 34–50.

Naish, M., Rawling, R. and Hart, C. (1987) *The Contribution of a Curriculum Project to 16–19 Education*. London: Longman/SCDC.

National Curriculum Council (1990a) *Consultation Report: Geography*. York: NCC.

National Curriculum Council (1990b) *Curriculum Guidance 3: The Whole Curriculum*. York: NCC.

National Curriculum Council (1990c) *Curriculum Guidance 4: Education for Economic and Industrial Understanding*. York: NCC.

National Curriculum Council (1990d) *Curriculum Guidance 5: Health Education*. York: NCC.

National Curriculum Council (1990e) *Curriculum Guidance 7: Environmental Education*. York: NCC.

National Curriculum Council (1990f) *Curriculum Guidance 8: Education for Citizenship*. York: NCC.

National Curriculum Council (1992) *Geography and Economic and Industrial Understanding at Key Stages 3 and 4*. York: NCC.

Neisworth, P., *et al.* (1969) *Student Motivation and Classroom Management: a Behavioristic Approach*. Lemont, Pennsylvania: Behavior Technics, Inc..

Newbigin, M. (1914) *The British Empire beyond the Seas: an Introduction to World Geography*. London: G. Bell and Sons.

Nixon, J. (1991) 'Reclaiming coherence: cross–curricular provision and the National Curriculum', *Journal of Curriculum Studies*, **23**, 187–92.

Northern Examinations and Assessment Board (1994) *A Comparability Study in GCSE Geography: A study based on the 1993 Examinations*. Manchester: NEAB.

Office for Standards in Education (OFSTED) (1993) *Geography: Key Stages 1, 2 and 3, Second Year, 1992–3: The Implementation of the Curricular Requirements of the Education Reform Act*. London: HMSO.

Orrell, K. (1994) 'Assessment at Key Stage 4 – emerging issues', in Walford, R. and Machon, P. (eds) *Challenging Times: Implementing the National Curriculum in Geography*. Cambridge: Cambridge Publishing Services. pp. 84–6.

Orrell, K. and Wilson, P. (1994) 'GCSE geography: the post–Dearing era', *Teaching Geography*, **19**, 90–1.

Parsons, C. (1987) *The Curriculum Change Game: a Longitudinal Study of the Schools Council Geography for the Young School Leaver Project*. Lewes: The Falmer Press.

Pattison, W. D. (1963) 'The four conditions of geography', in Bale, J., Graves, N.J. and Walford, R. (eds) (1973) *Perspectives on Geographical Education*. Edinburgh: Oliver and Boyd, pp.2–22.

Pearce, M. (1929) 'An individual method of instruction in the teaching of geography', *Geography*, **15**, 298–302.

Penstone, M. (1910) *Town Study: Suggestions for a Course of Lesson Preliminary to a Study of Civics*. London: National Society.

Peet, R. and Thrift, N. (eds) (1989) *New Models in Geography: the Political–Economy Perspective*. London: Unwin Hyman.

Pepper, D. (1993) *Eco–socialism: from Deep Ecology to Social Justice*. London: Routledge.

Peters, R.S. (1967) 'What is an educational process?', in Peters, R.S. (ed.) *The Concept of Education*. London: Routledge and Kegan Paul.

Pickles, T. (1932a) *Europe and Asia*. London: J.M. Dent and Sons (1959 edn.)

Pickles, T. (1932b) *Europe*. London: J.M. Dent and Sons (1945 edn.)

Pickles, T. (1939) *The World*. London: J.M. Dent and Sons.

Pike, G. and Selby, D. (1988) *Global Teacher, Global Learner*. London: Hodder and Stoughton.

Pollard, A. and Bourne, J. (1994) *Teaching and Learning in the Primary School*. London: Routledge, in association with the Open University.

Popham, W.J. (1968) 'Probing the validity of arguments against behavioural objectives', in Stones, E. and Anderson, D. (eds) (1972) *Educational Objectives and the Teaching of Educational Psychology*. London: Methuen, pp. 229–37.

Pring, R. (1971) 'Curriculum integration', in Peters, R.S. (ed.) (1973) *The Philosophy of Education*. Oxford: Oxford University Press, pp.123–49 .

Pring, R. (1989) *The New Curriculum*. London: Cassell.

Proctor, N. (1984) 'Geography and the common curriculum', *Geography*, **69**, 38–45.

Raths, J. (1967) 'Worthwhile activities', in Raths, J., Pancella, J.R. and Van Ness, J.S. (eds) (1971) *Studying Teaching*. Englewood Cliffs, NJ: Prentice Hall, Inc., pp.131–7.

Ravenstein, E.G. (1885) 'The aims and methods of geographical education', in Royal Geographical Society (1886) *Report of the Proceedings of the Royal Geographical Society in Reference to the Improvement of Geographical Education* (The Keltie Report). London: John Murray, pp.163–76.

Rawling, E. (1992) 'The making of a national geography curriculum', *Geography*, **77**, 292–309.

Rawling, E. (1993) 'School geography: towards 2000', *Geography*, **78**, 110–6.

Reade, W.W. (1863) *Savage Africa*. London.

Relph, E. (1981) *Rational Landscapes and Humanistic Geography*. London: Croom Helm.

Relph, E. (1989) 'Responsive methods, geographical imagination and the study of landscapes', in Kobayashi, A. and Mackenzie, S. (eds) *Remaking Human Geography*. Boston: Unwin Hyman, pp.149–163.

Reynolds, J.B. (1901) 'Class excursions in England and Wales', *The Geographical Teacher*, **1**, 32–6.

Reynolds, L. (1906) 'Education in the open–air', *The Geographical Teacher*, **3**, 152–9.

Rhys, W. (1972) 'The development of logical thinking', in Graves, N.J. (ed.) *New Movements in the Study and Teaching of Geography*. London: Temple Smith, pp.93–106.

Rich, P.B. (1986) *Race and Empire in British Politics*. Cambridge: Cambridge University Press.

Roberson, B.S. and Long, M. (1956) 'Sample studies: the development of a method', *Geography*, **41**, 248–59.

Robinson, R. (1994) 'Cross–curricular benefits for classroom geography', *Teaching Geography*, **19**, 178–9.

Roderick, G.W. and Stephens, M.D. (1978) *Education and Industry in the Nineteenth Century: the English Disease?* London: Longman.

Rogers, R., Viles, H. and Goudie, A. (eds) (1992) *The Student's Companion to Geography* Oxford: Blackwell.

Rosen, H. (1967) 'The language of textbooks', in Cashdan, A. and Grugeon, E. (eds) (1972) *Language in Education: a Source Book*. London: Routledge and Kegan Paul/Open University Press, pp. 119–25.

Rosen, H. (1969) 'Towards a language policy across the curriculum', in Barnes, D. (ed.) *Language, The Learner and The School*. Harmondsworth: Penguin, pp.119–168.

Royal Geographical Society (1886) *Report of the Proceedings of the Royal Geographical Society in Reference to the Improvement of Geographical Education* (The Keltie Report). London: John Murray.

Royal Geographical Society (1903) 'Syllabuses of instruction for geography', *The Teachers' Times*, **21 August**, 153–4.

Sandeman, G. (1915) 'The Iberian Peninsula', in Belloc, H. *et al.*, *Highroads to Geography: Book 4 – The Continent of Europe*. London: Nelson, pp.212–28.

Sargant, W.L. (1867) 'On the progress of elementary education', *Journal of the Statistical Society*, **30**, 80–137.

Sauer, C.O. (1925) 'The morphology of landscape', *University of California Publications in Geography*, **2**, 19–53.

Scarfe, N. (1969) 'Curriculum planning in geographic education', *New Zealand Journal of Geography*, **47**, 20.

School Board Chronicle (1900) 'Geographies and the Irish race', **17 March**.

School Examinations and Assessment Council (1993) *Pupils' Work Assessed: Geography KS3*. London: SEAC.

Sebba, J. (1991) *Planning for Geography for Pupils with Learning Difficulties*. Sheffield: The Geographical Association.

Sebba, J. (1995) *Geography for All*. London: David Fulton Publishers.

Selby, D. (1987) *Human Rights*. Cambridge: Cambridge University Press.

Shennan, M.K. (1991) *Teaching about Europe*. London: Cassell.

Shennan, M.K. and Lawrence, F. (1980) *The Planning of School Courses of European Studies*. London: The Historical Association.

Skeggs, B. (1991) 'Postmodernism: what is all the fuss about?', *British Journal of Sociology of Education*, **12**, 255–67.

Skilbeck, M. (1971) 'Preparing curriculum objectives', *The Vocational Aspect*, **23**, 2–3.

Skilbeck, M. (1976) 'Ideologies and Values', in Dale, S. (ed.) (1976) *Curriculum Design and Development*. Milton Keynes: Open University, Unit 3.

Skilbeck, M. (1984) *School-Based Curriculum Development*. London: Harper and Row .

Slater, F. (1982) *Learning through Geography*. London: Heinemann.

Smart, P. (1973) 'The concept of indoctrination', in Langford, G. and O'Connor, D.J. (eds) *New Essays in the Philosophy of Education*. London: Routledge and Kegan Paul, pp.3–46.

Smith, D.M. (1974) 'Who gets what *where*, and how: a welfare focus for geography', *Geography*, **59**, 289–97.

Smith, F. (1931) *A History of Elementary Education 1760–1902*. London: University of London Press.

Soja, E. W. (1989) *Postmodern Geographies: the Reassertion of Space in Critical Social Theory*. London: Verso.

Spencer, C., Blades, M. and Morsley, K. (1989) *The Child in the Physical Environment: the Development of Spatial Knowledge and Cognition*. Chichester: John Wiley and Sons.

Spink, H.M. and Brady, R.P. (1958) *The Southern Lands: New Ventures in Geography Series*. Huddersfield: Schofield and Sims.

Stamp, L.D. and Stamp, E.S. (1930ff.) *The New Age Geographies Series*. London: Longmans, Green and Co.

Stamp, L.D. (1939) *Geography for Today: Book III – North America and Asia*. London: Longmans, Green and Co. .

Stembridge, J.S. (1941) *The New Oxford Geographies: Book 1 – Life and Work at Home and Overseas*. Oxford: Clarendon Press .

Stoddard, L. (1920) *The Rising Tide of Colour against White World Supremacy*. London: Chapman and Hall.

Stoddart, D.R. (1965) 'Geography and the ecological approach: the ecosystem as a geographic principle and method', *Geography*, **50**, 242–51.

Stoddart, D.R. (ed.) (1981) *Geography, Ideology and Social Concern*. Oxford: Basil Blackwell.

Stoddart, D.R. (1986) *On Geography and its History*. Oxford: Basil Blackwell.

Stoddart, D.R. (1987) 'To claim the high ground: geography for the end of the century', *Transactions of the Institute of British Geographers*, **12**, New Series, 327–36.

Storm, M. (1983) 'Geographical development education: a metropolitan view', in Bale, J. (ed.) *The Third World*: Issues and Approaches. Sheffield: The Geographical Association.

Sturgeon, M.K. (1887) 'The teaching of elementary geography – a practical lesson, with models,' *Journal of the Manchester Geographical Society*, **3**, 83–95.

Taba, H. (1962) *Curriculum Development: Theory and Practice*. New York: Harcourt, Brace and World.

Taba, H. and Elkins, D. (1966) *Teaching Strategies for the Culturally Disadvantaged*. Chicago: Rand, McNally and Co..

Tann, C.S. (ed.) (1988) *Developing Topic Work in the Primary School*. Lewes: The Falmer Press.

Task Group on Assessment and Testing (TGAT) (1987) *National Curriculum: A Report*. London: DES.

Tate, T. (1860) *The Philosophy of Education: or the Principles and Practice of Teaching*. London: Longman, Green, Longman and Roberts.

Taylor, P.J. (1985) *Political Geography: World–Economy, Nation–State and Locality*. London: Longman.

The Teachers' Aid (1902a), **33**, **January**, 355.

The Teachers' Times (1902b) 'For the geography lesson: The Kaffir children', **12**, **February**, 129 .

The Teachers' Times (1904) 'Froebel's "Mother Songs" for students', **14 October**, 304.

Thomas, H.G. (1937) *Teaching Geography*. London: Ginn and Co.

Thomas, P.R. (1970) 'Education and the new geography,' in Bale, J., Graves, N.J. and Walford, R. (eds), *Perspectives on Geographical Education*. Edinburgh: Oliver and Boyd, pp. 67–76.

Thrift, N. (1992) 'Apocalypse soon, or why human geography is worth doing', in Rogers, R. *et al.* (eds) *The Student's Companion to Geography*. Oxford: Blackwell, pp. 8–12.

Tolley, H. and Orrell, K. (1977) *Geography 14–18: a Handbook for School–based Curriculum Development*. London: Macmillan.

Trimmer, S. (1801) *The Oeconomy of Charity*. London.

Tyler, R. (1949) *Basic Principles of Curriculum and Instruction.* Chicago: University of Chicago Press.

Tyler, R. (1964) 'Some persistent questions in the defining of objectives', in Stones and Anderson (eds) *Educational Objectives and Teaching of Educational Psychology.* London: Methuen, pp.179–87 .

Unstead, J.F. (1922) *The "Citizen of the World" Series: Book 1 – The British Isles of Today.* London: Sidgwick and Jackson.

Unstead, J.F. (1928) 'The primary geography schoolteacher – what should he know and be?', *Geography*, **14**, 315–22.

Unwin, T. (1992) *The Place of Geography.* London: Longman.

Vaughan, J.E. (1972) 'Aspects of teaching geography in England in the early nineteenth century', *Paedagogica Historica*, **12**, 128–47.

Vygotsky, L.S. (1934) *Thought and Language.* Cambridge, Mass.: MIT Press.

Walford, R. (1969) *Games in Geography.* London: Longman.

Walford, R. (ed.) (1973) *New Directions in Geography Teaching.* London: Longman.

Walford, R. (1986) 'Games and simulations', in Boardman, D. (ed.) *Handbook for Geography Teachers.* Sheffield: The Geographical Association, pp.79–84 .

Walford, R. (1989) 'On the frontier with the new model army: geography publishing from the 1960s to the 1990s', *Geography*, **74**, 308–20.

Walford, R. (1992) 'Creating a National Curriculum: a view from the inside', in Hill, D. (ed.) *International Perspectives on Geographical Education.* Boulder, Colorado: University of Colorado Department of Geography/ Rand McNally, pp.89–100.

Wall, W.D. (1968) *Adolescents in School and Society.* Slough: National Foundation for Educational Research.

Warwick, D. (ed.) (1973) *Integrated Studies in the Secondary School.* London: University of London Press.

Weiner, G. (1994) *Feminisms in Education: an Introduction.* Buckingham: Open University Press.

Welton, J. (1906) *Principles and Methods of Teaching.* London: University Tutorial Press.

Welpton, W.P. (1914) 'The educational outlook on geography', *The Geographical Teacher*, **7**, 291–7.

Wheeler, D.K. (1967) *Curriculum Process.* London: University of London Press.

White, J.P. (1967) 'Indoctrination', in Peters, R.S. (ed.) (1973) *The Concept of Education.* London: Routledge and Kegan Paul, pp. 177–91 .

White, J.P. (1982) *The Aims of Education Restated.* London: Routledge and Kegan Paul.

White, J.P. (1990) *Education and the Good Life: Beyond the National Curriculum.* London: Kogan Page/Institute of Education, University of London.

Wiegand, P. (1986) 'Values in geographical education', in Tomlinson, P. and Quinton, M. (eds) (1986) *Values across the Curriculum.* Lewes: The Falmer Press, pp. 51–76.

Wiener, M. (1981) *English Culture and the Decline of the Industrial Spirit.* Cambridge: Cambridge University Press.

Willard, E. (1826) 'Geography for Beginners', quoted in Whipple, G.M. (ed.) (1933) *The Thirty-Second Yearbook of the National Society for the Study of Education: The Teaching of Geography.* Chicago: The University of Chicago Press, p.10.

Williams, M. (ed.) (1976) *Geography and the Integrated Curriculum: a Reader.* London: Heinemann.

Williams, M. (ed.) (1981) *Language, Teaching and Learning: Geography.* London: Ward Lock Educational.

Williamson-Fein, J. (1988) 'Limits to geography: a feminist perspective', in Fien, J. and Gerber, R. (eds) (1988) *Teaching Geography for a Better World.* Edinburgh: Oliver and Boyd, pp. 104–16.

Wiseman, S. (1972–3) 'The educational obstacle race: factors that hinder pupil progress', *Educational Research*, **15**, 87–93.

Wolch, J. and Dear, M. (eds) (1989) *The Power of Geography: How Territory Shapes Social Life.* Boston: Unwin Hyman.

World Wildlife Fund (1993) *Education for the 21st Century: Lifelines Primary Supplement.* Godalming: WWF UK Publishing Unit.

Wright, D.R. (1983) '"Colourful South Africa"? an analysis of textbook images', *Multi-racial Education*, **10**, 27–36.

Wright, D.R. (1985a) 'In black and white: racist bias in textbooks', *Geographical Education*, **5**, 13–17.

Wright, D.R. (1985b) 'Are geography textbooks sexist?', *Teaching Geography*, **10**, 81–4.

Yoxall, J.H. (1891) *The Pupil Teacher's Geography.* London: Jarrold and Sons.

Index

aid agencies/relationships 164, 191
aims of education ix, 1–11, 74–76, 142, 150
  empathy as an aim 3, 125, 148, 150, 152, 211
  personal autonomy as an aim 3–4, 8
aesthetic activities 44
affective domain (*see* attitudes and values)
ageism (*see under* stereotyping)
American High School Geography Project 37
American studies 51
assessment (*see also under* National Curriculum) ix, 75, 99–118
  accountability/quality control function 63–4, 99–102, 104–105, 205
  assessment-led curriculum 99, 105–106, 169–170
  assessment objectives 63, 198–199
  certification function 103
  competency-based 62, 117, 205
  course-work 110–13, 201
  criterion-referenced 100, 102, 118
  formative/monitoring/feedback-feedforward 63, 74–76, 96–98, 101–104, 115, 118
  good practice in 39, 96, 101–103, 105–118, 178
  ipsative 105
  moderation 10, 111, 180
  norm-referenced 102, 118
  motivational function/raising standards 100, 104
  oral 114
  predictive function 103–4
  profiles/records of achievement 115–7, 205
  purposes of assessment 103–105
  range of assessment techniques 63, 106, 189
  reliability 105–107, 112, 118, 178
  self-assessment 92, 105
  summative 101–103, 118
  validity 105, 107–109, 112, 118, 178, 180
attitudes (and values) ix, 5–11, 16, 18, 24–25, 49, 72, 77, 93–94, 119–120, 136–153, 199
  assessment of 114–115

Board of Education 33, 50
Boer War 124, 158

Brandt Commission 22
British Social Hygiene Council 158

capitalism 5, 19, 21–22, 142, 147, 151, 158, 161
cartoons (*see* visual stereotyping)
Centre for World Development Education 142
child-centred education 4, 9, 42, 44–46
citizenship education 8, 125, 130, 155–156, 163–164, 190, 215
cognitive domain (*see* taxonomy of curriculum planning by objectives – taxonomy)
cognitive (mental) mapping 16, 78–79
colonialism (*see* imperialism)
concentric approach 32, 37–38, 55, 79, 81–82, 97, 145, 175
conceptual/principle development/learning 64–74, 78–83, 95
continuity
  primary/secondary 45, 182
  secondary/tertiary 18
Council for Education in World Citizenship 193
Council for Europe 165
cross-curricularity (integration) 8–9, 20, 34–35, 41–58, 70, 72, 140–145, 150, 154–166, 168
  areas of experience framework 129, 154–155
  dimensions 155, 165, 190–193
  skills 190, 209
  themes 154–155, 159, 161–165, 190
curriculum planning (*see also* National Curriculum Planning) viii, 3, 18, 39, 52, 77, 154
  concept-based schemes 67–74, 149–150, 180
    key concepts 20, 68–70, 144
    key ideas (principles) 56–57, 68, 72–4, 145–146, 183
    key questions 11, 144, 173, 183, 188, 191
  frameworks/models 74–76, 146
  by objectives (rational curriculum planning) 59–67, 74–77, 144, 152
    behavioural objectives 61–63
    process objectives 62–64
    taxonomies of objectives 64–67, 137–139
curriculum theory viii–ix, 2, 38–39, 59–77, 170, 215
  structure of knowledge 48–49, 51, 64, 67

Department for Education (DFE) 58, 171, 192–193, 209

228

Department of Education and Science
(DES) 9, 167–168
Department of Trade and Industry 161
determinism 29, 31, 124–125, 127,
132–133, 139–140
developed countries 147, 159, 173, 185
developing countries 22, 140, 144–148,
159, 173, 185, 191
Development Education Centre 142, 161,
190–191
differentiation (matching) 81, 98, 168,
183, 188–189, 197, 201, 204
distant places 139

economic and industrial understanding
150, 157, 161–162, 205
anti-industrial spirit 9, 156–157
EATE Project 161
national economic needs 9, 156–157
eco-socialism 23, 211
Education Act (1870) 155, 158
Education Act (1986) 5, 142
Education Reform Act (1988) vii, 4, 99,
118
educational worthwhileness (see good
practice)
empathy (see under aims)
enquiry-based learning 41, 75–77, 92–93,
113, 143, 151, 159, 169, 172–173,
183–184, 199, 202, 207–208
environmental education 49, 51, 55, 142,
162–163, 169, 190, 214–215
environmental geography 15–16, 20, 141,
151–153, 162–163, 173–174, 210
eugenics (see health education)
European Community/Union 57, 130,
174–175, 199, 201
European dimension/studies 51, 56–57,
134, 146, 160, 185, 190–194
examinations, GCE, CSE, GCSE, etc. 34,
51, 89, 91, 94, 100, 103, 105–106,
195–205
differential practice of examining
boards 201–202
eleven plus 117, 139
grade descriptions 202–205
modular syllabuses 202, 205
National Criteria 196–201
Schools Council syllabuses 40

field-work 31–34, 39, 55, 110–112,
168–169, 172–173, 177, 198–200
flexible/individualised learning 91–93,
96–99, 110, 183–184, 188
games and simulations 151–153
gender influence (see under progression)
geographers 125
Geographical Association 36, 42, 50, 86,

115, 125, 168, 171, 181, 196, 209,
214–215
geographical 'cube' (linking places, skills
and themes) 172, 177, 181–182, 185
geographical education
capes and bays geography 28–29, 31,
36, 45, 130–132, 136, 213
enlightened traditionalism 31–32,
36–37, 39
hard-core traditionalism 28–29, 36
sample (case) studies 34–36, 39,
133–134
geographical educationists 18, 38–39, 55,
125, 150, 212–213
geographical resources 186–188
maps as resources 171, 186–187
Ordnance Survey maps 79, 173
photographs as resources 72, 79, 87, 97,
112, 136, 145, 171, 173, 187
Geography 37, 168
geography as a curricular pivot 49–51,
169
globalising function 8, 22, 25–26, 51,
55, 137, 162–163, 171, 173, 175,
184–185, 189–190, 194, 198–199, 210
links with
area studies 28, 51, 55–56
art 50
basics 50
environmental subjects 51
history 50
humanities/social studies 49–51, 72,
134, 140
outdoor pursuits 51
sciences 49–50, 113
world studies/global studies/peace
studies/development education 51,
141–146, 163
geography as a discipline 12–27
behavioural/phenomenological
geography 15–16, 19–20, 38, 140, 151
critical/radical 10, 19–24, 27, 140, 142,
210–211
distinctiveness/place study 12–13, 20,
22, 24, 27–28, 36, 74, 136, 145,
168–169, 172–175, 177–178, 185–186,
210
feminist 23–24, 128–129, 211
humanistic 19–21, 27, 140, 150, 152,
211–212
mappability 12–13
Marxist/socialist 19–23, 211
political economy 22, 25
quantitative/scientific revolution 13–18,
37–41, 50, 210, 213
regional 13–15, 27–28, 30–31, 36–37,
59, 174–175, 185–186, 213
systematic geography/geographical

themes 13–14, 28, 174, 178, 185
welfare/human rights/peace studies 5, 10, 16–17, 21, 41, 129, 139–142, 145, 150–152, 163, 205, 210–211, 214
traditions 13, 18–19, 28, 74
geography 14–16 ix, 195–205
good practice in
education/pedagogy vii, ix, 8, 31,59, 62, 74, 76–77, 92, 97–98, 167, 180, 184, 188–189, 205, 212, 215
geography 59, 74, 92, 167, 180, 183–188, 205, 212–213
social education 59, 74, 159, 180, 183–184, 189–194, 205, 212–214

graphicacy 79, 87, 97, 113, 116, 169, 173, 198–199

health education 142, 157–158, 162
Herbartarians (type-studies) 34–35
HMI (*see also* OFSTED) 45, 99, 167–168
Geography 5–16 Curriculum Matters series 159, 165, 167, 180, 188–189
Reports/Surveys 45, 167–168
hermeneutics 20–21

ideologies/ideologues
educational 3, 8–11, 42, 44–45, 58, 65, 91, 118, 164
in geography 18–19, 128, 140, 143, 211
imperialism/nationalism 28–29, 121, 123–127, 130–132, 135–136, 142–143, 146, 155–156, 159, 165, 194
indoctrination 4–5, 142, 155, 157
information technology 87, 92, 110, 113, 168, 170, 172–173, 190, 193, 208–210
integration (*see* cross-curricularity)
Inter-departmental Committee on Physical Deterioration (1904) 158
interdependence 159, 163, 175, 191, 194
inter-(multi-)(trans-)disciplinary enquiry 48–49, 51–56, 70, 140–141, 164, 215
problems of
balance 51–52
focus 52
sequence 52–55
International Geographical Union (Charter on Geographical Education) 166, 213–214
International Monetary Fund 22
international understanding (*see also* citizenship education) 159–166, 177, 194, 214
issues
based ix, 49, 57–58, 136, 139–141, 145, 148, 151, 158, 202, 212
dominated 141, 143, 148
permeated 58, 70, 129, 141, 143,

149–152
language in geography 83–85, 98
abstract/concrete 69–71, 80, 180
technical/vernacular 69–71, 80, 84–85
League of Nations 125
Liverpool School Board 124
local education authorities (LEAs) 117, 142, 208, 212, 215
locality studies (see also sample/case studies) 34–35, 177, 213
contrasting 97
local 22, 32, 35, 55, 139, 177
locational knowledge 29, 169, 198, 207
cultural literacy 71

Manchester Geographical Society 123
maps (*see* graphicacy and geographical resources)
mastery learning (reinforcing success) 95–96
matching (*see* differentiation)
meritocracy 100, 155
moral education/dispositions/instruction/ training 3–8, 131, 148–149
multi-cultural/multi-ethnic education (*see under* cross-curricular dimensions)

National Curriculum vii–ix, 1–2, 10, 18, 27, 36, 53, 58, 62, 81, 86, 91, 93–94, 97–100, 102, 116–117, 142, 159, 164–167, 195–205, 207–208, 213
assessment
Attainment Targets/Statements of Attainment 99, 102, 116, 178, 181–183, 185, 196
Level Descriptions 175–180, 183, 185, 204
Dearing review ix, 53, 117, 155, 160, 163, 166–167, 171–180, 182–183, 195, 207–208
Level Descriptions (*see* assessment)
Geography Working Group 18, 159–160, 168, 172, 177, 182, 185, 195
Welsh Statutory Orders 175
National Curriculum Council 86, 102, 149, 151, 154–155, 159–164, 195–196
National Curriculum planning ix, 99, 154–166, 181–194
National Health Service 162
Northern Partnership for Records of Achievement 117

OFSTED 170–171, 181–182, 184, 205

personal and social education 165
Plowden Report 139
political education 159, 161
politicisation of the curriculum viii, 154–156, 158–159, 161, 167, 206

post-modernism 17, 24–27, 190, 211
Primary Geographer 168
Progression 52, 81–83, 175, 183, 188–189
  factors influencing
    ability 85–86
    ethnic minority/culturally
    disadvantaged groups 94–95, 109
    gender 87, 97, 109, 112, 210
    motivation 3, 64, 88–89, 94–95, 98,
    110
    reinforcing success/mastery learning
    95–96
    special needs 86–87
    time 88

racism (*see under* stereotyping)
religious instruction 4, 28, 155, 157
resources (*see* geographical resources)
Revised Code/Codes of the 1860s 99–100,
  117, 158
right-wing ideology/press/pressure groups
  vii, 4–5, 10, 142–143, 155, 169, 206
Royal Geographical Society 29, 33, 123

sample studies (*see under* geographical
  education)
Schools Council 39–41, 56, 68, 75, 90,
  212
  Avery Hill 14–16 Project (GYSL) 40,
  91, 141, 200, 202
  Bristol 14–18 Project 38, 40, 200, 202
  Liverpool 8–13 Project 40, 46, 138, 144
  London 16–19 Project 40, 143
Schools Curriculum and Assessment
  Authority (SCAA) 171, 174–175, 177,
  180, 185, 198–199, 204
Schools Examinations and Assessment
  Council (SEAC) 102, 197–198
school links (twinning) 142, 163, 193, 209
  Town Twinning Association 194
Secretaries of State for Education 99, 109,
  155, 159–160, 168, 195–196
sexism (*see under* stereotyping)
Social Darwinism 121–122, 124, 157
spatial cognition (*see* cognitive maps)
special needs 86–87, 98
spiral curriculum 52, 54–55, 81–83, 98,
  170, 180, 183, 189
stereotyping ix, 9, 19, 35–36, 43, 45, 57,
  119–136, 148, 156–157, 206, 213
  ageism 129–130
  check-lists 144–145, 190–191
  gender/sexism 126–129, 132, 144, 157
  racism 115, 120–126, 128, 131–133,
  139, 144–145
  visual (caricatures/cartoons) 120,
  134–136, 145
subject-based approaches 9, 42–45, 58, 70,
95, 140

Task Group on Assessment and Testing
  (TGAT) 99–103, 195
teacher education
  initial 207–208, 212
  INSET 170, 212, 215
teachers
  classroom management 90–91, 98
  craft 1–2, 61, 207
  deficit view of children 136, 179
  expectations/self-fulfilling prophecy 40,
  44–45, 79, 83, 90, 111
  generalist class teachers 43–44, 168
  influence 89–90
  professional ix, 1–2, 101, 111, 164, 166,
  170, 205–208
  questioning skills 84, 93–94, 188–189
  role ix, 57–58
  styles 90–91, 98
Teaching Geography 168
theory and practice of education 1–2
topic/thematic frameworks 43–44, 47, 49,
  52–55, 112, 144–145, 165, 215
trans-national corporations 22, 146–151,
  165, 191

UNESCO 164–166
United Nations Organisation (UNO) 164,
  166
Universal Declaration of Human Rights
  150, 165–166
utilitarianism/vocationalism 9, 155–157,
  161, 168
  GNVQ 196, 205
  TVEI 92, 205
values (*see* attitudes)

World Bank 22
World Health Organisation 162
World Wildlife Fund 214